The Readable Darwin
The Origin of Species
Edited for Modern Readers

The Readable Darwin

The Origin of Species

Edited for Modern Readers
by Jan A. Pechenik

Chapters 1 to 8 from the 1872 Sixth Edition

Sinauer Associates, Inc., Publishers
Sunderland, Massachusetts U.S.A.

About the Cover The cover was conceived and designed by Ardea Thurston-Shaine, and realized by Jefferson Johnson. Colorization of Darwin portrait by Dave Davis, neitshade5.wordpress.com; laptop photo © karandaev/istock; pigeon photo © sekernas/istock.

The Readable Darwin: *The Origin of Species*, **Edited for Modern Readers**

Sinauer Associates, Inc.
P.O. Box 407, 23 Plumtree Road
Sunderland, MA 01375 U.S.A.
Phone: 413-549-4300
FAX: 413-549-1118
email: orders@sinauer.com
www.sinauer.com

Library of Congress Cataloging-in-Publication Data

Darwin, Charles, 1809-1882.
  [On the origin of species. Chapters 1-8]
  The readable Darwin : the origin of species edited for modern readers :
chapters 1 to 8 from the 1872 sixth edition / Jan A. Pechenik.
      pages cm
  Includes bibliographical references and index.
  ISBN 978-1-60535-328-9 (pbk.)
  1. Evolution (Biology) 2. Natural selection. I. Pechenik, Jan A. II. Title.
  QH365.O25P43 2014
  576.8'2--dc23
                    2018017462

Printed in the U.S.A.

Contents

8   Instinct   223

Preface by Jan A. Pechenik

"You care for nothing but shooting, dogs, and rat-catching, and you will be a disgrace to yourself and all your family." (From *Charles and Emma*, 2009 by Deborah Heiligman.)

Charles Darwin received this rant from his father, Robert Darwin (a successful medical doctor and investor), as a teenager while studying medicine at the university in Edinburgh, in Scotland. He had just decided to abandon a career in medicine after seeing several people being operated on without anesthesia.

So the father's outburst is perhaps understandable. But how wrong he was! Here are some comments about the importance of Charles Darwin's original *The Origin of Species*:

"It's safe to say that *The Origin of Species* …is one of the most influential books ever written." David Quammen, *The Reluctant Mr. Darwin: An Intimate Portrait of Charles Darwin and the Making of His Theory of Evolution*. 2006. W. W. Norton & Company.

The Origin "is a book that makes the whole world vibrate." Adam Gopnik, *Angels and Ages: Lincoln, Darwin, and the Birth of the Modern Age*. 2009. Vintage Press.

"The Darwinian revolution is widely considered the most important event in the entire history of the human intellect." Michael Ghiselin. 2008.

Charles Darwin published *The Origin of Species* in 1859, one year after he and Alfred Russel Wallace had their remarkably similar papers on natural selection presented at a meeting of a major scientific society, the Linnean Society, in London, England. I think it's fair to say that the world hasn't been the same since. Here is the crux of his argument:

- Organisms vary in a great many characteristics (traits), both anatomical and behavioral.

- These characteristics are typically passed on to offspring—they are inherited.

- Organisms compete for space and food and mates, and are subject to predation, so that there is a constant struggle to stay alive, and to successfully reproduce.

- Those pressures give an advantage to individuals with certain characteristics.

- Individuals with those beneficial traits are most likely to survive and leave offspring, which will then inherit those useful traits.

- Over very long periods of time those characteristics will come to be found in a great many individuals, and a new species will be thereby created.

- Diversification and extinction will then, over incredibly long periods of time, create species that differ more and more from each other, creating new classes, families, genera...even new phyla.

It's a brilliant argument, and the argument is brilliantly and exhaustively made.

The first printing of *The Origin* sold out in the first day. Remarkably though, the book is rarely read today, not even in high school or college biology courses. That's a shame. Not only is it the basis for so much of our modern biological research in so many areas, but it's also a wonderful reminder of the incredible diversity of life on this planet, and a wonderful example of a fair and honest argument based on evidence and logical thinking. Darwin explains his supporting evidence thoroughly, but is also very up-front about the things he doesn't yet understand, even those that might pose problems for his theory.

The book *is* a bit of a slog. The style of writing is very much of the 1800's, and although some of his prose was delightfully memorable, many of his paragraphs are long and unwieldy, as are many of the sentences, making the reading rather difficult for many modern readers. He also refers to many people without ever saying who they are, and mentions the names of a great many animals and plants that most people today are not familiar with.

To make it easier for teachers and students to read this wonderful and important book, and to make it more accessible as well for interested adults, I have now edited the first 8 chapters into more readable prose. In doing so, I have tried to keep as much of Darwin's original wording as possible, and to preserve all of the sense of what he is saying.

I have chosen to work with Darwin's final edition (the Sixth), which was published in 1872. The First Edition was published in 1859, and was revised as new information was collected (by Darwin and by many others around the world) and as new arguments were advanced against Darwin's proposition. So in the Sixth Edition we can see how Darwin responded to all of his critics. For the Sixth Edition Darwin also added an entirely new and wonderful chapter (Chapter 7), which includes fascinating information on the evolution of baleen whales from toothed ancestors, the evolution of climbing in plants, and the evolution of breasts in mammals. And it was in the Sixth Edition of *The Origin* that Darwin introduced the word "evolution" (he had first used it in *The Descent of Man*, published the previous year).

One disadvantage of reading the Sixth Edition rather than the first, however, is that as Darwin was under increasing pressure to develop a complete theory of natural selection, he did more grasping at straws than he did in the first edition in dealing with the possible causes of variability among individuals and the inheritance of those variations by offspring. Even by the time of Darwin's death, in 1882, nobody knew what caused variation among individuals or how those variations were transmitted to offspring. Mendel published what eventually turned out to be the beginnings of modern genetics in 1866, but the paper was cited only 3 times in the next 35 years. It was published in an obscure journal, and those who read the paper seem not to have understood the implications, particularly that the findings applied to all sorts of traits, not just the especially distinctive traits that Mendel had been working with in his pea plant studies. Intriguingly, Mendel read a German translation of Darwin's work, but as far as we know he never tried to contact Darwin, even when he visited England in 1862. Apparently Mendel himself didn't see the connection between Darwin's ideas and his own work. In any event, it was many decades later that the implications of Mendel's work were realized, and several decades after that that the basic mechanisms of inheritance, including the role of DNA, were finally worked out and the mechanism of evolution through the process of natural selection, as Darwin had proposed, was finally well accepted. Indeed, some of that work is still going on.

So Darwin didn't get everything right; I have included a number of footnotes to indicate where Darwin was correct, and where he was wrong. But his ideas about the role of variation in shaping morphological and behavioral diversity through natural selection were on target, as were his ideas about the role of natural selection in gradually creating new species and new larger taxonomic categories. As a result, natural selection is now widely accepted as the major explanation for present and past organismal diversity, and for the appearance of major new groups of organisms over time.

My Edits

I have spent much time during the last 3–4 years editing the first 8 chapters, making multiple passes over every paragraph. But this is *not* my book: it's Darwin's, and I have tried to maintain his voice as much as possible. I have based much of my editing on the various rules from my *Short Guide to Writing About Biology* (Eighth Edition). Indeed, I like to think that this is how Darwin would have written *The Origin* had he read my *Short Guide* first! Here are the main rules that I have followed while editing this material, along with some examples:

1. Omit unnecessary words, especially unnecessary prepositions.

    Original: ...I suspect that the chief use of the nutriment in the seed is to favor the growth of the young seedling... (Chapter 3)

    My revision: ...I suspect that the seed's nutrients are chiefly used to promote the growth of the young seedling...

2. Eliminate weak verbs (what I call "Wimpy Verb Syndrome"), in part by making organisms the agent of the action.

    Original: Many plants are known which regularly produce...

    My Revision: Many plants regularly produce...

3. Warn readers of what lies ahead, using words like "For example," "Similarly," and "On the other hand...," and remind readers from time to time of what they have just read.

4. Incorporate definitions into sentences.

    Original: One of the most serious is that of **neuter insects**, which are often differently constructed from either the males or fertile females (Chapter 6, p. 174)

    My revision: One of the most serious is that of **neuter, non-reproductive insects**, which are often differently constructed from either the males or the fertile females of the species.

5. Use repetition, summary, and appropriate punctuation to link thoughts and improve the flow of ideas: Never make readers back up!

    Original: Although many statements may be found in works on natural history **to this effect**, I cannot find even one that seems to me of any weight.

    My revision: Although many works on natural history **claim that some structures in one species do indeed serve for the exclusive benefit of a different species**, I cannot find even one example that seems to me to hold any weight.

I have made a variety of other changes as well. One would-be reader commented on Amazon.com that "Another thing which made the book a little harder for me is Darwin mentions a lot of people and animals I've never heard of." I now clarify what the various animals and plants are that Darwin refers to. For example, when Darwin mentions cirripedes or *Balanus*, I help readers by including the term "barnacles" somewhere in the sentence.

I also have added a little more information to some sentences, to help make them more meaningful to modern readers. For example:

Original: Thus, we can hardly believe that the webbed feet of the upland goose or of the frigate-bird are of special use to these birds.

My revision: Thus, we can hardly believe that the webbed feet of the upland goose of South American grasslands, or of the frigate-bird, which cannot swim or even walk well, and which takes most of its food in flight, are of special use to these birds.

I have also broken up overly long sentences and overly long paragraphs, and altered sentence structure when necessary, to make the sentences easier to read.

Original: A trailing palm tree in the Malay Archipelago climbs the loftiest trees by the aid of exquisitely constructed hook clusters around the ends of its branches, and this contrivance, no doubt, is of the highest service to the plant; but as we see nearly similar hooks on many trees which are not climbers, and which, as there is reason to believe from the distribution of the thorn-bearing species in Africa and South America, serve as a defense against browsing quadrupeds, so the spikes on the palm may at first have been developed for this function, and subsequently have been improved and taken advantage of by the plant, as it underwent further modification and became a climber.

My revision: In my version, that single sentence has now become 3 sentences.

Original: Seeing how important **an organ of locomotion the tail is** in most aquatic animals....

My revision: Seeing how important **the tail is as an organ of locomotion** in most aquatic animals....

Finally, I have written one-paragraph previews for the start of each chapter, have bold-faced particularly important sentences, and have added many illustrations. Darwin's Sixth Edition originally had only a single illustration, a schematic drawing of evolutionary branching patterns, in Chapter 4. My version includes numerous drawings and photographs of the various plants and animals that Darwin talks about, along with some other figures that clarify some of his major points.

If enough people read this version of *The Origin* with enthusiasm, I will soon get to work on editing the remaining chapters and produce a second volume of *The Readable Darwin*. Note that in this volume, I refer several times to chapters in the second half of the original book, as did Darwin. Impatient readers might want to try reading those chapters as Darwin wrote them.

Acknowledgments

Preparing this "translation" gave me a wonderful opportunity to read a variety of books about *The Origin* and about Darwin's life. In particular, James Costa's *The Annotated Origin* (2009, Harvard University Press) and David Reznick's *The Origin Then and Now—an Interpretative Guide to the Origin of Species* (2010, Princeton University Press) were invaluable in helping me to fully understand Darwin's thinking.

I also wish to thank my friend and colleague Gordon Hendler at the Natural History Museum of Los Angeles County for critical feedback on Chapter 3 and for moral support, advice, and encouragement throughout the project. My colleague George Ellmore also provided helpful information about plant biology.

Thanks to the following reviewers for their critiques of the book proposal and sample chapter:

Stevan Arnold, Oregon State University
John Avise, University of California, Irvine
Alex Badyev; University of Arizona
David Begun, UC, Davis
Douglas Futuyma, The State University of New York at Stony Brook
Richard Harrison, Cornell University
Mark Kirkpatrick, University of Texas at Austin
Carol Lee, University of Wisconsin
Michael Lynch, Indiana University
Mohamed Noor, Duke University
Patrick Phillips, University of Oregon
Adam Porter, University of Massachusetts
David Rand; Brown University
Michael Wade, Indiana University
Bruce Walsh, University of Arizona

I am also grateful to the following biologists who read and commented on complete chapters of the manuscript:

Greg Bole, The University of British Columbia
Becky Fuller, University of Illinois at Urbana-Champaign
Friða Jóhannesdóttir, Cornell University
Adi Livnat, Virginia Polytechnic Institute and State University
Joel W. McGlothlin, Virginia Polytechnic Institute and State
Benjamin Normark, University of Massachusetts Amherst
Jeremy Searle, Cornell University
Christopher S. Willett, University of North Carolina at Chapel Hill
Pamela Yeh, University of California, Los Angeles
Rebecca Zufall, University of Houston University

It's also a pleasure to thank Andy Sinauer for his enthusiastic reception of this idea and his willingness to see it through, and to his entire team for their professionalism and expertise in bringing the project to completion so smoothly: Dean Scudder, Christopher Small, Chelsea Holabird, Stephanie Bonner, David McIntyre, Joan Gemme, Jefferson Johnson, and Tom Friedmann.

My son Oliver and his wife Ardea Thurston-Shaine helped me find many appropriate photographs, answered many of my questions about birds, and offered helpful suggestions on various drafts. Ardea also contributed an excellent drawing of various pigeon breeds for Chapter 1, and developed the concept that resulted in the fabulous image that adorns the cover of this book. One of my graduate students (Casey Diederich) and several undergraduates (Elizabeth Card and Chinami Michaels) also contributed wonderful figures for the project. And finally I thank my entire family—especially my wife, my Emma—for their love, support, and patience throughout the adventure. J. S. Bach, Franz Schubert, and Carl Nielsen have also played major roles in keeping me sane.

JAN PECHENIK
2014

Introduction

For nearly five years, from Dec. 27, 1831, until Oct. 2, 1836, I served as naturalist aboard the H.M.S. *Beagle*, exploring. During that voyage I was much amazed by how the various types of organisms were distributed around South America, and how the animals and plants presently living on that continent are related to those found only as fossils in the geological record elsewhere. These facts, as will be seen in later chapters, seemed to me to throw some light on the origin of species—that "mystery of mysteries," as it has been called by one of our greatest scientists, John Herschel.[1] After I returned home, it occurred to me in 1837 that I might be able to help address this great question by patiently accumulating and reflecting on all sorts of facts that might have any bearing on it. Finally, after five years of work, I allowed myself to speculate on the subject and wrote up some brief notes. I enlarged these in 1844 into a sketch of the conclusions that seemed to be most probable from the evidence I had collected. Over the subsequent 15 years I have steadily pursued the same object: trying to understand how new species come about. I hope you will excuse me for entering these personal details of my work, as I give them only to show that I have not been hasty in coming to a decision.

Now, in 1859, my work is nearly finished.[2] Still, it will take me many more years to complete it, and as my health is not strong, I have been urged to publish this brief version of my findings. I have more especially been urged to do this as Mr. Alfred Russel Wallace, who is now studying the natural history of the Malay Archipelago north of Australia, has reached almost exactly the same conclusions that I have reached concerning the origin of species. In 1858 he sent me an account on the subject, requesting that I pass it along to the geologist Sir Charles Lyell, who in turn sent it for presentation at a meeting of the Linnean Society. Mr. Wallace's paper has now been published, in the

---

[1] John Herschel was a chemist, astronomer, mathematician, botanist, inventor, and the son of the brilliant astronomer William Herschel.

[2] The first edition of *The Origin of Species* was published in 1859. Subsequent editions were published in 1860, 1861, 1866, 1869, and 1872, as Darwin collected more information and refined and expanded his ideas.

third volume of that society's journal. Sir Lyell and my colleague the excellent botanist Dr. Joseph Hooker, who both knew of my work—Dr. Hooker having read my sketch of it in 1844—honored me by suggesting that I also publish a summary of my ideas in that journal (some brief extracts from my own manuscripts) along with Mr. Wallace's excellent paper.[3]

The brief summary of my ideas that I record here must necessarily be imperfect. I do not have sufficient space in this volume to give references and authorities for many of my statements; I must trust the reader to have some confidence in my accuracy. No doubt errors will have crept in, though I hope I have always been cautious in trusting to good authorities alone. I can here give only the general conclusions at which I have arrived, with a few facts in illustration, which I hope, in most cases, will suffice. No one can feel more clearly than I do the necessity of eventually publishing in detail all the facts, with references, on which I base my conclusions, as I am well aware that some evidence can be found apparently leading to conclusions directly opposite to those that I have arrived at, for nearly every point I make. A fair result can be obtained only by fully stating and balancing the facts and arguments on both sides of each question; and it is impossible to do so here.

I also much regret that lack of space prevents my having the satisfaction of acknowledging the generous assistance I have received from many naturalists, some of whom I have never actually met. I cannot, however, let this opportunity pass without expressing my deep obligations to Dr. Hooker, who, for the last 15 years, has aided me in every possible way with his large stores of knowledge and his excellent judgment.

In considering the origin of species, it is certainly conceivable that some naturalist, just by reflecting on the mutual interactions of all living organisms, on their embryological similarities and differences, their geographical distribution, their geological succession over time, and other such facts, might reach the conclusion that individual species had not been independently created but had in fact descended from other species. But such a conclusion, even if well founded and well argued, would be unsatisfactory until it could be shown *how* the innumerable species inhabiting this world had been modified so as to acquire that perfection of structure and coadaptation that now justly excites our admiration. What is the mechanism by which this could come about?

Naturalists continually refer to external conditions, such as climate and food, as the only possible source of variation. In one limited sense, as I will discuss in detail later, this may be true; but it is preposterous to attribute to mere external conditions the structure of any organism—for instance that of the woodpecker, with its feet, tail, beak, and tongue so wonderfully adapted to catch insects under the bark of trees. And consider too the case of the mistletoe plant, which draws its nourishment from particular trees, and which has seeds that must be transported by certain birds, and which has

---

[3] For more information about this historic joint presentation on July 1, 1858, on the topic of evolution by natural selection, see Link 2 at the end of this chapter.

flowers with separate sexes absolutely requiring the help of certain insects to bring pollen from one flower to the other. It is equally preposterous to account for the structure of this parasitic plant, and its intimate and essential relationship with such a range of other organisms, by the effects of external conditions, or of habit, or of the wishes and desires of the plant itself.

Thus is it of the highest importance to gain a clear insight into the means through which organisms become modified and coadapted to interact with other organisms. When I began my observations it seemed to me that a careful study of domesticated animals and of cultivated plants would offer me the best chance of resolving this difficult problem. Nor have I been disappointed: in this and in all other perplexing cases, I have invariably found that our knowledge of variation under domestication, imperfect as it is, provided the best and most satisfying clues to how the process works in the wild. I am fully convinced of the value of such studies, even though they have so far been typically neglected by naturalists.

For this reason I devote the first chapter of this summary of my ideas and findings to Variation under Domestication. We shall see that a large amount of hereditary modification is at least possible, and perhaps even more importantly, we shall see how great is our power to accumulate slight variations in traits over time, by simply choosing which animals to breed together.

In Chapter 2, I will talk about the variability of species in the wild, although I'll not have space here to present the long catalogs of facts that the topic really requires. I will, however, be able to discuss the circumstances that are most favorable to causing variation. In the next chapter (Chapter 3), I will consider the struggle for existence that takes place among all living things throughout the world, and show how it is an inevitable consequence of the exponential rates[4] at which the populations of all living beings tend to grow (see Figure 3.3[5]). This is the basic doctrine of the English cleric and demographer Robert Malthus, applied now to the whole animal and vegetable kingdoms: **As many more individuals of each species are born than can possibly survive, and as, consequently, there is a frequently recurring struggle for existence, it follows that any organism will have a better chance of surviving if it varies even just slightly in any way that is helpful to itself under the complex and sometimes varying conditions of life.** Thus, that individual will be *naturally selected for*. As so many variations are transmitted through inheritance, any selected variety will then tend to propagate its new and modified form among its offspring.

I will discuss this fundamental topic of "natural selection" in some detail in Chapter 4, and will show how natural selection almost always causes extinction of the less capable forms of life and leads to what I have called "divergence of character" within a population. In the next chapter

---

[4] With exponential growth, a population continues to grow by the same percentage each year, which means that the population size will increase faster and faster over time.
[5] Figure 3.3 was not included in Darwin's *The Origin of Species*.

(Chapter 5) I will discuss the complex and little-known laws of variation. In the succeeding three chapters, I will carefully consider the most apparent and gravest difficulties presented by my theory. These include 1) the difficulty in understanding how a simple organism or a simple organ can eventually be changed and perfected into a highly developed being or into an elaborately constructed organ (Chapters 6 and 7); and 2) the difficulty in understanding how complex animal behaviors and mental capacity may be shaped by the same forces that shape morphology (Chapter 8).[6]

No one should be surprised that there is as yet much that is unexplained regarding the origin of species and varieties, considering how profoundly ignorant we are about the mutual interrelations of all the animals and plants that live around us. Who can explain why one species ranges widely and is abundant, while a related species has a narrow range and is rare? Yet these relationships are of the greatest importance, for they determine the present welfare, and, I believe, the future success and modification of every inhabitant of this world. We know even less of the mutual relations of the innumerable inhabitants of the world during the many past geological epochs in Earth's history. **Although much remains obscure, and will long remain obscure, I can entertain no doubt, after the most careful study and dispassionate judgment of which I am capable, that the view which most naturalists entertain, and which I previously entertained myself—that each species was independently created—is erroneous.** I am fully convinced that species are not unchangeable, but rather that all those species that now belong to what are called the same "genera" are direct descendants of some other species, a species that is probably now extinct. In the same way, the acknowledged varieties of any one species are the direct descendants of that species. Furthermore, I am convinced that natural selection has been the main engine, although not the exclusive one, of modification over long periods of time.

Online Resources *available at* **sites.sinauer.com/readabledarwin**

Links

Link 1 With the following link, you can see all of the changes that Darwin made for each of the six editions of *The Origin of Species*.

http://darwin-online.org.uk/Variorum

Link 2 For more information about the historic Wallace-Darwin joint presentation concerning the origin of species by means of natural selection, on July 1, 1858, see the following link:

http://en.wikipedia.org/wiki/On_the_Tendency_of_Species_to_form_Varieties;_and_on_the_Perpetuation_of_Varieties_and_Species_by_Natural_Means_of_Selection

(*Note: Web addresses may change. Go to sites.sinauer.com/readabledarwin for up-to-date links.*)

---

[6] *The Origin of Species*, Sixth Edition, contains 15 chapters; *The Readable Darwin* covers Chapters 1–8.

1

Variation Under Domestication

In this chapter, Darwin talks about how domesticated animals and plants have gradually come to look as they do through our powers of selecting for the particular traits that we value, over many generations. The key to this process is individual variability, and the passing of particular traits to offspring.

Variability[1]

For any given species, when we compare members belonging to the same variety or subvariety of our older cultivated animals and plants—there are at least 7,500 varieties of apples, for example, all members of the species *Malus domestica*—one of the first things we notice is that they generally differ more from each other than do the individuals of any one species or variety in nature. It seems clear that organisms must be exposed during several generations to new conditions to cause any great amount of variation; but once the characteristics have begun to vary, they generally continue to vary for many generations. We know of no case in which any variable organism has ceased to vary under domestic cultivation. Our oldest cultivated plants, such as wheat, still yield new varieties, and our oldest domesticated animals can still be rapidly modified or improved, because of the variations that they continue to exhibit.

Effects of Habit and of the Use or Disuse of Parts; Correlated Variation; Inheritance

Changed habits produce an inherited effect, as, for example, in when during the year a plant begins to flower after being transported from one climate to

---

[1] I have shortened this section considerably. As you will see throughout this book, Darwin, having no knowledge of chromosomes, DNA, or even the basics of Mendelian genetics (first put forward by Gregor Mendel, in an obscure paper published in 1866), was at a loss to explain the causes of variation, or how those variations were transmitted to offspring. He returns to this topic in Chapters 2 and 5.

another. With animals, the increased use or disuse of parts has had a more marked influence. Thus I find that in the domestic duck, the bones of the wing weigh less in proportion to the weight of the entire skeleton than do the same bones in the wild duck, while the bones of the leg weigh more; this change may be safely attributed to the domestic duck flying much less, and walking more, than its wild parents in nature.[2]

Many laws regulate variation, some few of which can be dimly seen. Let me first say something about what may be called "correlated variation." Major changes in embryonic or larval development, for example, will probably cause some changes in the mature animal. Some instances of correlation are quite whimsical: for example, cats that are entirely white and have blue eyes are generally deaf, while the work of Dr. Karl Friedrich Heusinger von Waldegg suggests that while white sheep and pigs are poisoned by certain plants, dark-colored individuals are not. The American biologist Professor Jeffries Wyman has recently given a good illustration of this phenomenon: when he asked some Virginia farmers why all of their pigs were black, they told him that the pigs ate the paint-root plant (genus *Lachnanthes*), which colored their bones pink and caused the hoofs of all but the black individuals to fall off, so that the farmers now select only the black members of each litter for raising and breeding, since those have the best chance of surviving. Hairless dogs have imperfect teeth; long-haired and coarse-haired animals have long horns or many horns; pigeons with feathered feet also have skin between their outer toes; pigeons with short beaks have small feet, and those with long beaks have large feet; and animal breeders believe that long limbs are almost always accompanied by an elongated head. Thus if humans go on selecting, and thus augmenting, any peculiarity in a species, we will almost certainly modify other parts of the structure—completely unintentionally—owing to the mysterious laws of correlation.

The causes of variation are unknown, or at best only dimly understood, and the results of those laws are infinitely complex and diversified. Similarly, the laws governing the inheritance of those variations are for the most part unknown: No one can say why the same peculiarity in different individuals of the same species, or in different species, is sometimes inherited and sometimes not; or why a child often reverts in certain characteristics to its grandfather or grandmother, or even to a more remote ancestor; or why a peculiarity is often transmitted from one sex to both sexes, or to one sex alone, more commonly but not exclusively to a member of the same sex.

But the key issues here are that both plants and animals do vary in many characteristics within a species, and that many of those variations are passed along to offspring. Indeed, the number and diversity of inheritable

---

[2] In the absence of any convincing data to the contrary, Darwin thought that the ideas of Jean-Baptiste Lamarck (1744–1829), a prominent French naturalist, about the inheritance of changes resulting from the increased use or disuse of various body parts seemed reasonable. As mentioned earlier, Darwin knew nothing about the causes of variation or the laws governing inheritance.

variations in structure, of both slight and great physiological importance, are endless. It is those inherited variations, regardless of what the causes of variation are or what the mechanisms of inheritance turn out to be, that are important to my argument.

Character of Domestic Varieties; Difficulties of Distinguishing between Varieties and Species; Origin of Domestic Varieties from One or More Species

Members of the various domesticated races often differ from each other, and from other species within the same genus, to an extreme degree in some particular part, especially when compared with the species in nature to which they are most closely allied. With these exceptions, domesticated races belonging to a single species differ from each other to about the same degree that closely allied species of the same genus differ from each other in nature. Thus, the distinction between species and the varieties found within a single species is not easily made. Indeed, the domesticated races of many animals and plants have been ranked by some competent judges as the descendants of what were originally several distinct species, and by other competent judges as mere varieties of a single species. If the distinction between species and domesticated races within species was markedly clear, this source of doubt would not be so common.

When trying to estimate the amount of structural difference between allied domestic races, we are soon placed in doubt by not knowing for sure whether the various races are descended from one or from several different parent species. This point, if it could be cleared up, would be quite interesting: if, for instance, it could be shown that the greyhound, bloodhound, terrier, spaniel, and bull-dog, which we all know propagate their kind very accurately, were in fact all offspring of a single ancestral species,[3] then we would surely doubt the immutability of the many closely allied natural species—for instance, of the many fox species—inhabiting different parts of the world. Unfortunately, in the case of our most anciently domesticated animals and plants, it is not possible to come to any definite conclusion about whether they are descended from one or from several wild species.

It has often been assumed that man has chosen to domesticate only animals and plants that have an extraordinarily developed natural tendency to vary, and likewise to withstand diverse climates. I do not dispute that these capacities have increased the value of most of our domesticated organisms. But how could a savage possibly know, when he first tamed some particular animal, whether it would vary in succeeding generations, and whether it would be able to tolerate other climates? Has the small amount of natural variability that we see in the donkey and the goose, or

---

[3] As shown by modern molecular work, this turns out to be the case, contrary to what Darwin believed: all dogs did in fact evolve from a single ancestral species, the gray wolf, *Canis lupis*.

Figure 1.1 *Gallus bankiva* now goes by *Gallus gallus bankiva*, the red junglefowl.

the limited ability of reindeer to withstand warmth, or the limited ability of the common camel to withstand cold prevented their domestication? No, it has not. I cannot doubt that if other animals and plants belonging to as great a diversity of classes and countries as our domesticated productions belong to were taken from nature and bred for an equal number of generations under domestication, they would on average vary just as much as the parent species of our existing domesticated productions have varied.

The origin of most of our domestic animals will probably remain vague, but some authors hold the absurd belief that every race that breeds true has had a separate prototype species in nature, even if the distinctive characteristics between the races are ever so slight. If so, then there must have existed at least 20 species of wild cattle and at least 20 species of wild sheep in Europe alone. Indeed, one author believes that there must formerly have been at least 11 wild species of sheep unique to Great Britain!

Although those beliefs seem extremely unlikely, I do believe that a small part of the difference between the various domestic dog breeds is due to their being descended from several distinct species. But in the case of strongly marked races of some other domesticated species, there is presumptive or even strong evidence that all are truly descended from a single wild stock.

For example, having kept nearly all the English breeds of fowl alive myself over some years, having bred and crossed them, and having examined their skeletons very carefully, it seems to me almost certain that all are descended from the wild Indian species *Gallus bankiva* (Figure 1.1). Indeed, this is exactly the conclusion reached by Mr. Edward Blyth and others who have studied this bird in India. Similarly for ducks and rabbits, some breeds of which differ a great deal from each other, the evidence is clear that they are all descended from the common wild duck and rabbit.

I would like to talk now about how we have made good use of inherited variations over the years in domesticating our animals and cultivating our crops.

Breeds of the Domestic Pigeon, Their Differences and Origin

Believing that it is always best to study some special group, I have, after much careful thought and investigation, taken up the breeding of domestic pigeons. One advantage of working with pigeons is that males and females

Figure 1.2 A sampling of pigeon diversity. From left to right: line 1, fantail, runt, English carrier; line 2, barb, rock pigeon, turbit; line 3, pouter, short-faced tumbler, Jacobin.

can be easily mated for life: thus different breeds can be kept together in the same aviary without fear of them mating randomly with each other. I have kept every breed of pigeon that I could purchase or otherwise obtain, and several people have kindly sent me pigeon skins from all around the world. Many works about pigeons have been published, in a variety of languages; some of these writings are especially important to us because of their great antiquity. I have also associated with several eminent pigeon breeders, and have been allowed to join two London Pigeon Clubs. So I have learned much about pigeons and about breeding them.

The diversity of pigeon breeds is truly astounding (Figure 1.2). Here I will talk about just a few of them: the English carrier pigeon; the short-faced tumbler; the "runt;" the "barb;" the pouter; the turbit; and the Jacobin, tumbler, and laugher pigeons.

First compare the English carrier pigeon with the short-faced tumbler, and see the wonderful difference in their beaks, which relates to corresponding differences in their skulls. The carrier pigeon, and especially the male, is also remarkable for the wonderful development of the fleshy outgrowth on the skin about its head; this is accompanied by greatly elongated eyelids, very large external openings to the nostrils, and a wide gaping mouth (Figure 1.3A).

In contrast, the so-called "runt" is a very large bird, with a long massive beak and large feet; some of the subspecies of runts also have very long necks, while others have very long wings. Some also have long tails, while others have singularly short tails.

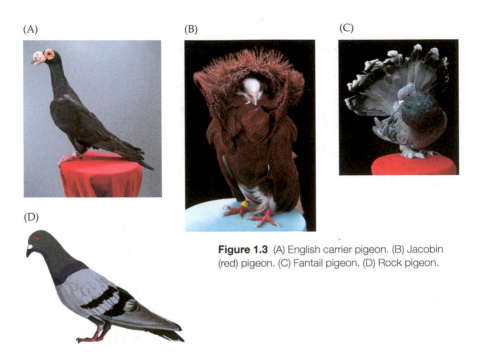

(A)

(B)

(C)

(D)

Figure 1.3 (A) English carrier pigeon. (B) Jacobin (red) pigeon. (C) Fantail pigeon. (D) Rock pigeon.

The bird known as the "barb" is similar to the carrier pigeon, except that instead of a long beak it has a very short and broad one. The turbit, on the other hand, has a short and conical beak, with a line of reversed feathers down its breast, and the characteristic habit of continually expanding the upper part of its esophagus slightly, while the Jacobin pigeon has its feathers so reversed along the back of its neck that they form a hood; it also has wing and tail feathers that are unusually long compared with the size of the bird (Figure 1.3B).

The pouter has a decidedly elongated body, wings, and tail, and its enormously large crop[4] (see Figure 1.2), which it glories in inflating, may well excite astonishment, and even laughter. The fantail pigeon has 30 or even 40 tail feathers, instead of the 12–14 found in all other pigeons, and it keeps those feathers expanded and erect so that in many individuals the head and tail touch (Figure 1.3C)! Also, the oil gland, common to all other pigeons, is completely absent in fantails.

Behaviors also differ among the breeds: For example, the trumpeter and laugher pigeons, as their names suggest, utter a very different-sounding "coo" from members of the other breeds, and the common tumbler pigeon has the very distinctive, inherited habit of flying at a great height in a compact flock, and then tumbling in the air, head over heels: No other pigeon does this! There are also several other less distinct breeds that I could talk

---

[4] A crop is an expandable, muscular pouch near the bird's throat, used to store food.

(A)

(B)

(C)

(D)

Figure 1.4 (A) Double-crested priest pigeon. (B) Modena pigeon. (C) Frillback pigeon. (D) Scandaroon pigeon.

about. But I think I have made my point about how much the behaviors of the various pigeon breeds differ from each other.

In the skeletons of the various pigeon breeds, the bones in the face vary enormously in length and width and degree of curvature; the shape of the lower jaw bone, as well as its length and width, also vary in an extremely remarkable manner. The caudal and sacral[5] vertebrae vary in number, as does the number of ribs, the width of the ribs, and whether or not the ribs bear outgrowths. The size and shape of the openings in the sternum vary greatly from breed to breed, as does the amount of divergence and relative size of the two arms of the wishbone (i.e., the furcula).

Many other structures also vary a great deal among the different pigeon breeds (Figure 1.4): The proportional width of the gape of the mouth; the proportional length of the eyelids, of the opening of the nostrils, and of the tongue (and not always in exact correlation with the length of the beak); the size of the crop and of the upper part of the esophagus; the degree of development of the oil gland, and even whether it is present or not; the number of primary wing and tail feathers; the length of the wing compared to the length of the tail and the length of the body; the relative lengths of the leg and foot; the number of plates on the toes; and the degree to which skin develops on the toes.

---

[5] Caudal vertebrae are those in the tail, for animals with tails; the sacrum is found in the center of the pelvis.

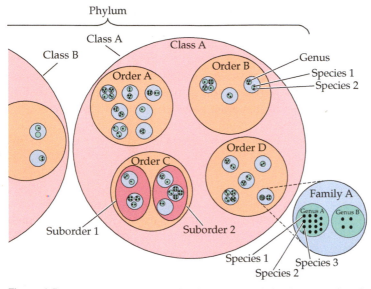

Figure 1.5 A general view of the relationship between phyla, classes, orders, families, genera, and species. Each genus may contain many species, with some genera containing more species than others. One family typically contains many genera, and some families contain more genera than others. Similarly, families are contained within distinct orders, orders are contained within distinct classes, and classes are contained within phyla. There are about 30 well-defined animal phyla. All vertebrates—including for example, dogs, cats, birds, frogs, fishes, turtles, whales, and people—are contained within the phylum Chordata.

And there's more. The age at which the birds acquire their final plumage also varies among breeds, as does the state of the down with which the nestling birds are first clothed when they hatch from the egg. The shape and size of the eggs also varies considerably among pigeon breeds, as does the manner in which the birds fly. In some pigeon breeds the voice and disposition also vary remarkably, as I mentioned earlier. Finally, the males and females have come to look slightly different from each other in some breeds but not others.

The different breeds of pigeon vary so greatly in appearance that at least a dozen different sorts of pigeons might be brought to any ornithologist,[6] and if he were told that they were wild birds rather than domestically-bred birds, he would certainly rank each of them as a well-defined, separate species. Moreover, I do not believe that any ornithologist would under such circumstances even place the English carrier pigeon, the short-faced tumbler, the runt, the barb, the pouter, and the fantail in the same genus! See Figure 1.5 for the distinction between species, genera, and other taxonomic categories. And yet, as great as the differences are between the breeds of pigeon, I am fully convinced that the common opinion of naturalists is correct: **all of these**

---

[6] An ornithologist is an expert on the biology of birds.

birds are descended from a single ancestor—the rock pigeon, *Columba livia* (see Figure 1.3D). Let me explain my reasoning.

If the various pigeon breeds I have just talked about are not varieties of a single species, and have not all descended from the rock pigeon, then they must have descended from at least seven or eight separate ancestral stocks; it would be impossible to make the present domestic breeds by cross-mating any of the present breeds with each other. How, for instance, could you get a pouter pigeon by crossing males and females of any other two breeds, unless one of the parent stocks possessed the characteristically enormous crop of the pouter? The supposed ancestral stocks must all have been rock pigeons. But none of the two or three other known rock pigeon species have any of the characteristics found among the domestic pigeon breeds. Thus the supposed ancestral pigeon stocks must either still exist in the countries where they were originally domesticated and yet be unknown to any ornithologist—which seems rather improbable, considering the size, habits, and remarkable characteristics of these birds—or they must all have become extinct in the wild, which is also most unlikely: birds breeding on precipices, and that are good fliers, are unlikely to go extinct. Indeed, the common rock pigeon, which has the same habits as the domestic breeds, has not been exterminated in the wild even on several of the smaller British islands, or on the shores of the Mediterranean. Thus the supposed extermination of so many bird species having similar habits with the rock pigeon seems a very rash, unlikely assumption.

Moreover, the several domesticated breeds that I have been talking about have been transported to all parts of the world, and some of them must also have been carried back again into their native country. Yet not one of them has become wild or feral,[7] even though the dovecot pigeon, which is the rock pigeon in only a very slightly altered state, has indeed become feral in several places. Also, all recent experience shows that it is difficult to get wild animals to breed freely when they are first domesticated. And yet, on the hypothesis that our domesticated pigeons had many separate origins from many different ancestors, we would have to assume that at least seven or eight species were so thoroughly domesticated in ancient times by half-civilized man as to reproduce quite prolifically under confinement. Again, that is not very likely.

Another argument against the idea of our present pigeon breeds having descended from many separate ancestors, and one that carries great weight, is the fact that the various breeds I have mentioned are very much like the rock pigeon in their constitution, habits, voice, coloring, and in most parts of their structure, and yet are highly *unlike* the rock pigeon in many other aspects. For example, we may look in vain through all the wild species in the family Columbidae[8] for a beak that looks like that of the English carrier pigeon, or that of the short-faced tumbler or the barb; or for reversed feathers like those

---

[7] Feral is a domesticated species that has reverted to living in the wild.

[8] This family contains the doves and pigeons, about 310 species. The Columbidae is one member of the class Aves, which includes all birds.

of the Jacobin; or for a crop like that of the pouter; or for tail feathers like those of the fantail. Thus it must be assumed that not only did half-civilized man succeed in thoroughly domesticating a number of very different pigeon species in ancient times, but that he intentionally or by chance picked out extraordinarily abnormal species to work with. And we must assume further that those very species have all since become either extinct or unknown. So many strange contingencies are extremely improbable.

Pigeon colors are also well worth thinking about. The rock pigeon is slaty blue in color, with white loins (see Figure 1.3D). The tail has a terminal dark bar, with the outer feathers externally edged in white at the base. The wings have two black bars. In some semi-domestic breeds, and in some truly wild breeds, the wings have not only the two black bars but are also checkered with black. These several marks do not occur together in any other species in the entire family.

Now, in every one of the domestic breeds, even in birds that have been carefully bred for many generations, all of the above marks that are seen in rock pigeons, even to the white edging of the outer tail feathers, some-times concur perfectly developed—not very often, but every now and again. Moreover, when birds belonging to two or more distinct breeds are crossed, none of which are blue or have any of the above-specified markings, the mixed offspring very often suddenly display those characters.[9] For example (and I could give many others), some time ago I crossed some white fantail pigeons, which breed very true, with some black barb pigeons. The offspring of this cross were black, brown, and mottled. I'll talk about what I did with these "mixed fantail—black-barb" offspring shortly. I also crossed a blue barb pigeon with a spot pigeon, which is a white bird with a red tail and a red spot on its forehead, and which is notorious for breeding very true to its type; in this case the "mixed barb—spot" offspring were somewhat dark ("dusky") and mottled, but certainly not blue. I then crossed a mixed fantail—black barb pigeon with a mixed barb—spot pigeon…and they pro-duced a bird of as beautiful a *blue* color as any wild rock pigeon! And it even had the rock pigeon's white loins, double black wing bars, and barred and white-edged tail feathers! Who would have predicted such a result? Blue varieties of barb pigeons are incredibly rare: indeed I have never heard of a single instance of blue-colored barb pigeons in all of England.

We can understand these remarkable outcomes quite easily on the well-known principle of "reversion to ancestral characters," if all of the domestic breeds are *indeed* descended from the rock pigeon, with its blue coloration, black bars, and white edging. But if we deny this ancestral relationship, we must make one of the two following suppositions:

1. Even though no other existing pigeons are now colored and marked like the rock pigeon, we could assume that all of the eight or more

---

[9] As used here, 'characters' means "traits."

imagined ancestral stocks of domestic pigeons were so-colored and so-marked, so that in each separate breed there might be a tendency to sometimes revert to the very same ancestral colors and markings, or

2. We could assume that each domestic breed, even the purest, has within the previous 12–20 generations been crossed with a rock pigeon. I say within 12–20 generations because we know of no instance in which any crossed descendants have ever reverted to an ancestor of foreign blood if they are more than 20 generations removed from that cross.

Both of these suppositions are highly unlikely.

Finally, consider the effects of crosses on future fertility, i.e., the ability of the offspring to reproduce successfully. Now it is very rare for matings between two distinct *species* to produce fertile offspring; typically the off-spring from such matings are sterile. And yet the hybrids resulting from crosses between any of the existing pigeon breeds are perfectly fertile, and have no difficulty in producing offspring themselves as adults; I can say this based on the results of my own studies, which were purposefully made using the most distinctly different-looking breeds available to me. Some authors believe that long-continued domestication eliminates the strong tendency toward sterility in matings between species. That conclusion may be correct if applied to species closely related to each other. But it would be rash in the extreme to extend that argument to suppose that separate species originally as distinct as carrier pigeons, tumblers, pouters, and fantails should now yield perfectly fertile offspring when crossed with each other.

In summary, we must consider the following issues:

1. The improbability of man having made seven or eight previously separate and distinct species of pigeons to breed freely under domestication;

2. These supposed seven to eight ancestral species being quite unknown in the wild, and the domesticated breeds not having become wild anywhere in nature;

3. These different breeds presenting a series of striking characteristics that are not found in any other members of the family Columbidae—which includes hundreds of species—even though they are so like the rock pigeon in most other respects;

4. The occasional reappearance of the blue color and various black marks of the rock pigeon in all of the breeds, both when kept pure and when crossed; and lastly,

5. The hybrid offspring from crosses between the different pigeon breeds being perfectly capable of reproducing and leaving their own offspring.

Taken together, we are lead to conclude that all of our domestic pigeon breeds are indeed descended from the rock pigeon, *Columba livia*. In support of this view, I may add that *C. livia*[10] taken from nature has been successfully domesticated in Europe and in India, and that it agrees in its habits and in a great number of morphological features with all the domestic breeds. Secondly, although the English carrier pigeon and the short-faced tumbler differ immensely from the rock pigeon in certain characteristics, yet by comparing the several sub-breeds of these two races, particularly those brought from distant countries, we can see an almost continuous series of traits between them and the rock pigeon. We can do this in some other cases as well, although admittedly not with all breeds. Thirdly, those characters that are mainly distinctive of each breed also vary *within* each breed; the wattle and length of beak of the carrier pigeon, for example, and the shortness of tumbler's beak, and the number of tail feathers in the fantail pigeon: all vary among individuals within each breed. The explanation of this fact—and its importance—will be made clearer later, when I talk about "selection."

Fourthly, pigeons have been watched and tended with the utmost care, and loved by many people, for thousands of years in several parts of the world; indeed, the earliest known record of pigeons is in the 5th Egyptian dynasty, about 3,000 B.C.E., as was pointed out to me by Professor Karl Richard Lepsius, the well-known German Egyptologist and author. But even better, Mr. Samuel Birch of the British Museum informs me that pigeons were also listed on a dinner menu from the previous dynasty! And Pliny the Elder (23–79 C.E.), in his *Naturalis Historia*, tells us that people paid immense prices for pigeons: "Many men are grown now to cast a special affection and love to these birds; they build turrets above the tops of their houses for dovecotes. Nay, they are come to this pass, that they can reckon up their pedigree and race, yea they can tell the very places from whence this or that pigeon first came."

Pigeons were also much valued by Akbar the Great in India, about the year 1600; at least 20,000 pigeons were taken wherever the court traveled. "The monarchs of Iran and Turan sent him some very rare birds," says the court historian, who continues: "His Majesty by crossing the breeds, which method was never practiced before, has improved them astonishingly." About this same time, the Dutch were as eager about pigeons as were the old Romans. So people have been raising pigeons for a very, very long time—more than 2,000 years. The paramount importance of these facts in explaining the immense amount of variation that pigeons have undergone will be obvious when I talk about "selection" later.

I have discussed the probable origin of our domestic pigeons at some length, because when I first kept pigeons at home and watched the several kinds, knowing well how truly they breed from one generation to the next, I found it just as difficult to believe that they had all descended from a common ancestor as any naturalist would in coming to a similar conclusion

---

[10] *C. livia* is the abbreviation for *Columba livia*. After using the full name once, it is standard practice to abbreviate the genus name thereafter in the same piece of writing.

about the many finch species, or other groups of birds, in nature.

One circumstance that has struck me in particular is that nearly all the domestic animal breeders that I have talked to—not just the breeders of pigeons—along with the cultivators of plants who I have spoken with or whose treatises I have read, are all firmly convinced that the several breeds with which each has worked are descended from just as many ancestrally distinct species: i.e., four breeds having four separate ancestors; eight breeds having eight separate ancestors. Ask, as I have asked, a celebrated raiser of Hereford cattle whether his cattle might not have descended from Longhorn cattle (Figure 1.6), or whether both may have descended from some common ancestral stock, and he

(A)

(B)

Figure 1.6 (A) Polled Hereford bull. (B) Texas Longhorn.

will laugh you to scorn. Similarly, I have never met a pigeon breeder, or a poultry, duck, or rabbit fancier, who was not fully convinced that each main breed was descended from a distinctly separate ancestral species. The Belgian scientist Professor Jean Baptiste Van Mons, in his treatise on pears and apples, shows how utterly he disbelieves that the several sorts of apple—a Ribston pippin apple or a codlin apple, for instance—could ever have resulted from the seeds of the same tree.

I could give innumerable other examples of such misleading opinions. The explanation for these opinions, I think, is simple: from long-continued study of their organisms these breeders have become strongly impressed with the *differences* between the several races, and though they well know that within each race the members vary slightly from each other—for breeders win their prizes by selecting such slight differences for show—yet they ignore all general arguments and refuse to sum up in their minds the effect of such slight differences accumulating over many successive generations. Now many naturalists know far less about the laws of inheritance than breeders have learned through practical experience, and know no more than breeders do about the intermediate links between varieties in the long lines of descent, and yet they freely admit than many of our domestic races have the same parents in common. May these naturalists not learn a lesson of similarly see an astonishing improvement in many florists' flowers simply by comparing the flowers of the present day with drawings made only 20 to 30 years ago. Once a race of plants is pretty well established, the seed raisers do not pick out the best plants, but merely go over their seedbeds and pull up the "rogues," as they call the plants that deviate from the proper standard. With animals this same sort of selection is, in fact, likewise followed; hardly anyone is so careless as to breed the next generation from his worst animals!

With plants, there is yet another way to observe the gradual accumulated effects of selection over many generations—namely by comparing the diversity of flowers in different varieties of the same species in a single flower garden; or the diversity of leaves, pods, or tubers, or whatever part of the plant is valued, in the kitchen garden in comparison with the flowers of the same varieties; or the diversity of the fruits produced by any one species in an orchard, in comparison with the leaves and flowers of the same varieties of that species. See how different the leaves of the cabbage are, but how extremely alike are the flowers. Notice how the flowers of the different varieties of heartsease (*Viola tricolor*) are so unlike each other, but how the leaves are so similar. See how much the fruits of the different kinds of gooseberries differ in size, color, shape, and hairiness, while the flowers themselves present only very slight differences. It's not that the varieties that differ largely in one feature do not differ at all in other features; this is hardly ever, and perhaps never the case—and I speak here after careful observation. As a general rule, it cannot be doubted that the continued selection of slight variations in particular features, either in the leaves, the flowers, or the fruit, will produce offspring differing from each other chiefly in those characteristics, and not in others.

Now some readers have objected that the principle of selection has been honed to a rigorous, methodical practice for scarcely more than 75 years. Well, yes, certainly the practice has become more common in recent years; many treatises have now been published on the subject. And yes, the result has been, in a corresponding degree, rapid and important. But the principle is certainly not a modern discovery. I could give several references to works of great antiquity, in which the full importance of the principle of selection is acknowledged. Long ago, in rude and barbarous periods of English history, choice animals were often imported, and in fact laws were passed to prevent their exportation; the destruction of horses smaller than a certain size was ordered, similar to the previously described "rouging" of plants by nurserymen.

Indeed, the principles of selection are clearly given in an ancient Chinese encyclopedia, and similarly explicit rules for selection are laid down by some of the Roman classical writers. From passages in Genesis, it is clear that people paid careful attention to the color of domestic animals even in those early days. Savages now sometimes cross their dogs with wild canine animals, to improve the breed, and they did so formerly as well, as attested to by ancient passages from Pliny the Elder. The primitive peoples of South caution, when they ridicule the idea of species in the natural world being direct descendants of other species?

Now we are ready to deal with a key question: How have pigeon breeds as different from each other as the pouter, the common tumbler, and the runt all been created from a single ancestor—the rock pigeon?

Principles of Selection Anciently Followed, and Their Effects

Let us now briefly consider the steps by which the various distinctive domestic races have been produced, either from one or from several related species. One of the most remarkable features in our domesticated races is how easily we see that they are adapted, not to the animal's or plant's own good, but to our own use or fancy. Some variations useful to us have probably arisen suddenly, in a single step. Many botanists, for instance, believe that the common weed known as Fuller's teasel (*Dipsacus fullonum*), with its specialized hooks that really cannot be rivaled by any mechanical contrivance that I know of, is simply a variety of the wild member of the same genus and may have changed to this degree suddenly, in a single seedling. So it has probably been with the turnspit dog, a dog bred for its long body and short legs. And this is known to be the case with the ancon sheep, which also have unusually long bodies and short legs, with the forelegs not only short but also crooked.

But when we compare the dray horse with the race horse, the dromedary with the camel, the various breeds of sheep suited for cultivated land or for mountain pasture; when we see that the wool of one breed of sheep is good for one purpose and that of another breed is good for a quite different purpose; when we compare the many breeds of dogs, each good for people in different ways; when we compare the gamecock, so pertinacious in battle, with other breeds that rarely quarrel, such as the bantam, which is so small and elegant; when we compare the host of agricultural, culinary, orchard, and flower garden races of plants, most useful to us at different seasons and for different purposes, or so beautiful in our eyes, we must, I think, look further than to mere variability. We cannot suppose that all the breeds were suddenly produced as perfect and as useful as we now see them, in a single step. Indeed, in many cases we know that this has not been their history. **The key to understanding these differences is in our power of accumulative selection: nature gives us successive variations, and we add them up over time in certain directions that are useful to us.** In this sense we may be said to have made these many useful breeds for ourselves.

The great power of this principle of selection is not hypothetical. Several of our most distinguished breeders have modified their breeds of cattle and sheep to a large extent, and within a single lifetime. To fully realize what they have done, you really need to read some of the many treatises devoted to this subject, and to inspect the animals personally. Breeders habitually speak of an animal's organization as something "plastic," something that they can mold almost as they please. If I had more space I could quote numerous passages to this effect from highly competent authorities. Take William Youatt, for example, who was probably better acquainted with the works of agriculturists than almost any other individual I know of, and who, being a well-respected veterinarian, was himself a very good judge of animals. He speaks of the principle of selection as "that which enables the agriculturist, not only to modify the character of his flock, but to change it altogether. It is the magician's wand, by means of which he may summon into life whatever form and mould he pleases." Similarly, Lord John Somerville, speaking of what breeders have done for sheep, says "It would seem as if they had chalked out upon a wall a form perfect in itself, and then had given it existence." In Saxony (a state in Germany) the importance of the principle of selection in regard to merino sheep is so fully recognized that men follow it as a trade: the living sheep are each placed on a table and are carefully studied, like a picture is studied by a connoisseur. This is done three times several months apart, and the sheep are each time marked and classed, so that only the very best individuals may ultimately be selected for breeding.

What English breeders have actually accomplished with their domesticated animals is proven by the enormous prices given for animals with a good pedigree; and these have been successfully exported to almost every quarter of the world. The improvements are by no means generally due to crossing different breeds with each other. And even when a cross has been made, the most careful selection among different traits is even more indispensable in that situation than it is when breeds have not been crossed. If selection consisted merely in separating some very distinct variety, and breeding from it, the principle would be so obvious as to hardly be worth noticing. **Instead, its importance lies in the great effect produced by the gradual accumulation in one direction, during successive generations, of differences so small as to be absolutely undetectable by an uneducated eye—differences that I myself have attempted to discern, but in vain.** Not one man in a thousand has the accuracy of eye and the judgment necessary to become a successful breeder. If someone is gifted with those qualities and studies his subject for years, and indeed devotes his lifetime to it with indomitable perseverance, he will succeed, and may make great improvements. But if he lacks any of those qualities, he will assuredly fail. Few people would readily believe in the natural capacity and the years of practice needed to become even a skillful pigeon fancier.

The same principles are followed by horticulturists, who work with plants, but the variations here are often more abrupt. No one supposes that our choicest plants have been produced by a single variation from the stock of some ancestral plant. Indeed we have proof that this has not been so in several cases in which exact records have been kept. To give but one small example, consider the steadily increasing size of the common gooseberry fruit, which is now some 800 percent heavier than it was in ancient times. We

Africa match their draught cattle by color, as some Eskimos do with their teams of dogs. The Scottish missionary David Livingstone states that good domestic breeds are highly valued by people living in the interior of Africa, people who have never before associated with Europeans.

Although some of these facts do not show actual selection, they certainly show that the breeding of domestic animals was carefully attended to even in ancient times, and is now attended to by even the most primitive peoples. It would, indeed, have been strange to learn that attention had not been paid to breeding, for the inheritance of good and bad qualities in offspring is so obvious.

Unconscious Selection

At the present time, eminent breeders try by methodical selection, generation after generation, to make a new strain or sub-breed that is superior to anything else that exists in the country, and they do so with a distinct object in view. But for our purpose, a form of selection that results from everyone trying to possess and breed only from the best individual animals—something we may call Unconscious Selection—is more important. Thus, of course, a man who intends to keep pointers tries to get the best dogs he can to start with, and afterwards breeds only from his own best dogs. But he has no wish or expectation of permanently altering the breed. Nevertheless we may infer that this process, continued during centuries, would improve and modify any breed, in the same way that Robert Bakewell and other early agriculturists did methodically modify using this very same process, the forms and qualities of their cattle even during their lifetimes. Slow and imperceptibly small changes of this kind can never be recognized unless actual measurements or careful drawings of the breeds in question were made long ago, to serve for comparison. In some cases, however, unchanged or slightly changed individuals of the same breed exist in less civilized districts, where the breed has been less improved.

There is reason to believe that the King Charles's spaniel has been unconsciously modified to a large extent since King Charles II ruled England, Scotland, and Ireland in the seventeenth century. Some highly competent authorities are convinced that another dog breed, the setter, is directly derived from the spaniel, and has probably been slowly altered from it. It is also widely recognized that the English pointer has been greatly changed within the last century; in this case the change has, it is believed, been brought about chiefly through deliberate matings with the foxhound. But the important point for us is that the changes have been brought about unconsciously and gradually, and yet so effectively that, although the old Spanish pointer certainly came from Spain, Mr. George Borrow tells me that in his travels he has not seen any native dog in Spain that resembles our pointer.

By a similar process of selection, and by careful training, English racehorses have come to surpass—both in fleetness and size—the parent stock of Arabian horses; indeed, the English horses, by the official regulations of the Goodwood Races in England, are now required to carry heavier weights than the Arabian racehorses, to make the races fairer. Similarly, the famous cattle breeder Lord Spencer and others have shown that English cattle are now heavier and mature earlier compared with the stock formerly kept in this country. And by comparing the accounts given in various old treatises of the former and present state of carrier and tumbler pigeons in Britain, India, and Persia, we can easily trace the stages through which they have insensibly passed and have thereby come to differ so greatly from their common ancestor, the rock pigeon.

The well-respected veterinarian William Youatt gives an excellent illustration of the effects of a course of selection in his treatise on sheep (*Sheep: their breeds, management, and diseases*, 1837), one that may be considered as "unconscious" in that the breeders could never have expected (or even wished) to produce the result that ensued—namely, the production of two distinct strains of sheep. As Mr. Youatt remarks, "The two flocks of Leicester sheep kept by Mr. John Buckley and Mr. Joseph Burgess have been purely bred from the original stock of Mr. Bakewell for more than 50 years. There is not a suspicion existing in the mind of any one at all acquainted with the subject, that the owner of either of them has deviated in any one instance from the pure blood of Mr. Bakewell's flock. And yet the difference between the sheep [now] possessed by these two gentlemen is so great that they have the appearances of being quite different varieties."

Even if somewhere there are savages so barbarous as never to think of the inherited character of the offspring of their domesticated animals, any one animal particularly useful to them in some way, for any special purpose, would nevertheless be carefully protected and fed during the famines and other incidents to which savages are so liable; those chosen animals would thus generally leave more offspring than the inferior ones, so that there would be a kind of unconscious selection going on over the generations. We see the great value set on non-human animals even by the barbarians of Tierra del Fuego, by their killing and devouring their old women in times of famine; the old women were clearly valued less than the dogs at such times.

In plants, we can see the same gradual process of improvement, again achieved quite simply through the occasional preservation of the best individuals, simply by noting the increased size and beauty that we now see in the varieties of the heartsease, rose, pelargonium, dahlia, and other plants, when compared with the older varieties or with their parent stocks (Figures 1.7 and 1.8).

Nobody would ever expect to get a first-rate heartsease or dahlia from the seeds of a wild plant. Similarly, nobody would expect to raise a first-rate melting pear from the seeds of the wild pear: indeed, the pear, though cultivated in classical times, appears from Pliny's description to have been a fruit of very inferior quality, and I have seen great surprise expressed in horticultural works at the wonderful skill of gardeners, in their having

(A)   (B)

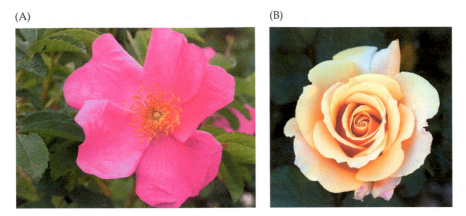

Figure 1.7 (A) Wild rose. (B) Domestic rose.

produced such splendid fruits from such poor starting materials. The art has been simple, and as far as the final result is concerned, has been followed almost unconsciously. It has consisted in always cultivating the best-known variety, sowing only its seeds, and, whenever a slightly better variety chanced to appear, selecting it, and so onwards over many generations. But the gardeners of the classical period, who always cultivated the best pears that they obtained, never thought what splendid fruit we should be eating today; and yet we owe our excellent fruit at least partly to their having naturally chosen and preserved the best varieties they could find in every generation.

A large amount of change, thus slowly and unconsciously accumulated, explains, I believe, the well-known fact that in a number of cases we cannot recognize (and therefore do not know), the wild parent stocks of the plants that have been cultivated in our flower and kitchen gardens

(A)   (B)

Figure 1.8 (A) Wild dahlia (*Dahlia sorensenii*). (B) A dahlia star sunset flower.

for the longest times. If it has taken centuries or even thousands of years to improve or modify most of our plants up to their present standards of usefulness to people, we can understand how it is that neither Australia, the Cape of Good Hope, nor any other region inhabited by uncivilized man has afforded us a single plant worth culturing. It is not that these countries, so rich in species, do not by strange chance possess the ancestral stocks of any useful plants; rather, it is simply that the native plants have not been improved by continued selection up to a standard of perfection comparable with that acquired by the plants in countries that were civilized long ago.

Domestic animals kept by uncivilized tribes almost always have to struggle for their own food, at least during some times of year. And in two countries very differently circumstanced, some individuals of the same species, differing slightly in their physiology or structure, would often succeed better in the one country than in the other; two sub-breeds might then eventually be formed by a process of "natural selection," as I will explain more fully later in this book. This, perhaps, partly explains why the varieties kept by primitive peoples are more like true species than are the varieties kept in civilized countries, something previously noted by a number of other authors.

Once we understand the important part that selection by people has played, then it becomes obvious how it is that our domestic races of animals and plants now show structural and behavioral adaptations to human wants and fancies. We can, I think, further understand the frequently abnormal characters of our domestic races and why they differ so greatly in external characters but so relatively slightly in their internal parts or organs. People cannot select, or can select only with much difficulty, for structural differences that are internal and cannot be seen; indeed, people rarely care for what is only internal. And of course we can only select variations that are first given to us in some slight degree by nature. Nobody would ever try to make a fantail pigeon until he saw a pigeon with a tail developed in some slight degree in an unusual manner. No one would try to create a pouter until he first saw a pigeon with a crop of somewhat unusual size; and the more abnormal or unusual any character was when it first appeared, the more likely it would be to catch his attention. But to use an expression like "trying to make a fantail" is, I have no doubt, in most cases utterly incorrect. The person who first selected a pigeon with a slightly larger tail never dreamed what the descendants of that pigeon would eventually look like through many generations of selection, partly unconscious and partly methodical. Perhaps the parent bird of all fantails had only 14 tail feathers somewhat expanded, like the present Java fantail, or was more like individuals of other distinct breeds, in which as many as 17 tail feathers have been counted. Perhaps the first pouter pigeon did not inflate its crop much more than the turbit now inflates the upper part of its esophagus—a habit of turbots that is ignored by all fanciers, as it is not one of the points of interest in that breed.

Remember, it does not take a great deviation of structure to catch the fancier's eye: as I noted earlier, the fancier can perceive extremely small differences. And it is part of human nature to value any novelty in one's possession, however slight. Nor must we assume that the value which breeders formerly saw in any slight difference among individuals of the same species be judged by the value that we now set on those differences, once distinct breeds have been firmly established. With pigeons, many slight variations still occasionally appear, but these are now typically rejected as faults or undesirable deviations from the standard of "perfection" that now characterizes that breed.

These views appear to explain why we know hardly anything at all about the origin or history of any of our domestic breeds. But, in fact, a breed, just like the dialect of a language, can hardly be said to *have* a distinct origin. Someone preserves and breeds from an individual with some slight but interesting deviation of structure, or takes more care than usual in matching his best animals and thus improves them. The improved animals then mate, and slowly their offspring spread in the immediate neighborhood. But they will as yet hardly have a distinct name, and from being so little valued, no one will have paid any attention to their history. When further improved over time, however, by the same slow and gradual process, they will spread more widely and will eventually be recognized by someone as being distinct and valuable, and will then probably first receive a local name. In semi-civilized countries, with little communication over long distances, the spreading of a new sub-breed would be a slow process. As soon as the points of value are once acknowledged, however, the principle that I am calling "unconscious selection" will always tend to slowly add to the characteristic features of the breed, whatever they may be. But there will be an infinitely small chance that anyone would have kept records of such slow, varying, and insensible changes.

Circumstances Favorable to Man's Power of Selection

Let me now say a few words about the circumstances that favor—and those that stand in the way of—man's power of selection. A high degree of variability in characteristics among individuals is obviously favorable, as it freely provides the materials for selection to work on. Even small individual differences are sufficient, though, with extreme care, to allow large modifications to eventually be accumulated in almost any desired direction. But having large numbers of individuals in the population is also important: Because variations manifestly useful or pleasing to us appear only occasionally, they are much more likely to appear when keeping a large number of individuals in that population. When only a few individuals are available, all will be allowed to breed, whatever their quality may be, and this will effectively prevent selection from taking place. In keeping with this idea, William Marshall has written with respect to Yorkshire sheep, in his *The*

Rural Economy of Yorkshire (1796), "As they generally belong to poor people, and are *mostly in small lots*, they never can be improved." On the other hand, nurserymen, from keeping large numbers of the same plant, are generally far more successful than amateurs in raising new and valuable varieties.

A large number of individuals of any particular animal or plant species can be reared only where the conditions favor its propagation. But probably the most important element in promoting selection is that the animal or plant should be so highly valued by man that he pays the closest attention to even the slightest deviations in its qualities or structure. Unless such attention be paid, nothing can be achieved. I have seen it gravely remarked that it was "most fortunate" that the strawberry began to vary just when gardeners began attending to this plant. But I'm sure that the strawberry has in fact varied ever since it was first cultivated; it's just that the slightest variations had for a long time been neglected. As soon, however, as gardeners picked out individual plants with slightly larger, earlier, or better fruit, and raised seedlings from those particular plants and again picked out the best seedlings and bred again from them, and so on, then (with some aid by crossing distinct species) those many admirable varieties of the strawberry were raised that have appeared during the last 50 years.

With animals, the ease of preventing unwanted crosses is also an important element in forming desirable new races—at least in a country that is already stocked with other races. In this respect, being able to enclose the land is helpful. In contrast, wandering savages or the inhabitants of open plains rarely possess more than one breed of the same species, because all members of the species can freely mate and breed. In this regard, pigeons are unusually easy to work with, since they can be mated for life. Thus many distinct races of pigeons may be improved and kept true even with all of the pigeons housed together in the same aviary, as mentioned earlier; this must have greatly aided the formation of new breeds by pigeon fanciers over the decades. Pigeons, I might add, can be propagated in great numbers and very quickly, and inferior individuals may be freely rejected...and eaten!

In contrast, cats, from their nocturnal rambling habits, cannot be selectively mated very easily and, even though they are much valued by women and children, we rarely see a distinct breed long kept up. The distinct breeds that we do sometimes see are almost always imported from some other country, often from islands. Although I do not doubt that some domestic animals vary in their characteristics less than others, yet the rarity or absence of distinct breeds of the cat, the donkey, peacock, goose, and others may be attributed mainly to selection not having been brought into play. For cats, the lack of selection is due to the difficulty of pairing them, as previously noted. For donkeys there is the problem that poor people keep only a few animals at a time, and pay little attention to their breeding; in contrast, this animal has been surprisingly modified and improved by careful selection in certain parts of Spain and the United States, showing that modifications are indeed possible. For peacocks, the problem is that they are not very easily reared, and a large stock of peacocks is not kept. With geese, the problem is that they are valuable for only two purposes—for food and for their feathers—and especially that there has been no pleasure felt in displaying distinct breeds.

Some authors have claimed that the maximum amount of variation in our domesticated animals and plants is soon reached, and can never afterwards be exceeded. However, it would be somewhat rash to claim that the limit of variation has been attained in any one case; for almost all of our animals and plants have been greatly improved in many ways within the recent past, which implies variation. It would be equally rash to claim that characters now increased to their utmost limit could not, even after remaining fixed for many centuries, begin to vary again under a changed environment. No doubt, as Mr. Alfred Russel Wallace has noted, a limit to at least some traits must eventually be reached. For instance, there must be a limit to the fleetness of any terrestrial animal, as this will be determined by the amount of friction to be overcome, the weight of the body to be carried, and the power with which the muscle fibers can contract. But what concerns us here is that the varieties of a given domesticated species differ more from each other in almost every character that man has attended to and selected for, than do distinct species within a single genus.

Isidore Geoffroy Saint-Hilaire has proved this in regard to size, and so it is with color and probably also with the length of hair. With respect to fleetness, which depends on many bodily characters, the undefeated English racehorse Eclipse was far fleeter, and a powerful dray horse is incomparably stronger—even though all horses belong to a single species—than are the members of any two natural species belonging to the same genus. And so it is with plants: the seeds of the different varieties of bean or corn plants probably differ more in size than do those from distinct species within any genus in the same two families (see Figure 1.5). This observation also applies to the fruit of the several varieties of the plum, for example, and still more strongly to seeds of the melon.

Summary

Let me now sum up my thoughts on the origin of today's domestic races of animals and plants. Variability among individuals in traits of interest to us is an essential ingredient, and those desired traits must be inherited by offspring. Variability among individuals is governed by many unknown laws, including correlated growth, environmental effects, increased use or disuse of particular parts, and in some cases the intercrossing of what were originally distinct species. But whatever the causes of that variability, the accumulative action of *selection*—whether applied methodically and quickly, or unconsciously and slowly (but efficiently) over many, many years—seems to have been the predominant power in creating the great variety of domesticated animals and plants that we have with us today.

Key Issues to Talk and Write About

1. Find out two interesting things about one of the people that Darwin mentions in this chapter. Choose from the following:

   Pliny the Elder
   Jeffries Wyman
   Robert Bakewell
   Isidore Geoffroy Saint-Hilaire
   Alfred Russel Wallace

2. Based on your reading of this chapter, what facts would you use in trying to convince someone that the different breeds of pigeon were really separate species?

3. What facts would Darwin use in trying to convince someone that the pouter, tumbler, fantail, and runt pigeons are all highly modified descendants of the rock pigeon?

4. How does Darwin explain the fact that some body parts of related animals or the fruits of some related plants are extremely different from each other, while other parts are extremely alike?

5. Suppose you wished to create a breed of pigeon that resembled a pouter, but with much larger toes than the pouters we have today. According to Darwin, how would you try to create such a modification?

6. Why is variation in traits among individuals of a species such an important part of Darwin's argument?

7. Figure 1.5 shows the relationship between phyla, classes, orders, families, genera, and species. Make a table comparing how humans and pigeons are classified into these categories. Just to get you started: both are in the animal phylum Chordata.

8. Why is Darwin so preoccupied in this chapter with convincing us that variations can be inherited?

9. Species today are generally defined by their reproductive isolation from other species; i.e., by their inability to mate with members of other species. How does Darwin seem to be defining species in this chapter?

10. Read the paragraph (page 13) that begins "If the various pigeon breeds...". Now write a one-sentence summary of that paragraph. First list what seem to be the several major points that Darwin is making. Then try to get all of those points into a single sentence. The sentence should be accurate, contain all the important points, and be in your own words, as though you were explaining something to a friend. It should also make perfect sense to a reader who has never read the original paragraph. It can be done!

2

Online Resources *available at* sites.sinauer.com/readabledarwin

Videos

1.1 Tumbler pigeon tumbling in flight

Links

1.1 A very nice summary of Lamarck's ideas about the causes of evolutionary change.

en.wikipedia.org/wiki/Lamarckism

1.2 A list of all of the American Kennel Club recognized dog breeds, including photos.

www.akc.org/breeds/breeds_a.cfm

1.3 The orange pippin website lists many of the 7,500 apple varieties presently in existence around the world.

www.orangepippin.com/apples

(Note: Web addresses may change. Go to sites.sinauer.com/readabledarwin for up-to-date links.)

Bibliography

Marshall, W. 1796. *The Rural Economy of Yorkshire.* London.

Youatt, W. 1837. *Sheep: Their Breeds, Management, and Diseases.* London.

2

Variation in Nature

In this chapter, Darwin emphasizes that characteristics vary greatly among individuals of all species, and gives many examples showing that there are often no clear boundaries between species and the varieties within a species. Deciding whether to call something a separate species or merely a variety within a species can be very subjective. He shows also that the most common and wide-ranging species tend to show the most variation among individuals. Variation is also greater for species belonging to larger genera (i.e., those containing many species) in any particular location than to those belonging to smaller genera. Also, many of the species in the larger genera resemble varieties, in having restricted ranges and in being rather similar looking. These facts make sense only if varieties within species gradually become distinct new species, and if species continue to generate new varieties over time.

In the previous chapter I talked about how people have gradually molded organisms in marvelous ways by selecting for small differences in particular traits over many, many generations. Before discussing how these same principles can apply to organisms in nature, I will first discuss the extent to which organisms in nature exhibit variation in traits. And indeed they do, and not only between species, but within species as well.

First, what are "species"? And what are "varieties"? And what do we mean by the term "variation"? Although no one definition of "species" has yet satisfied all naturalists, the term generally includes the unknown element of a distant act of creation. It's also hard to define the term "variety" precisely, but we generally assume that varieties are organisms that are closely related to each other but nevertheless differ in some conspicuous characteristics. Some authors have used the term "variation" to imply modifications that are directly due to changes in the physical environment and that are presumably not inherited by the offspring of those individuals. But who can say whether the unusually small shells of snails living in the brackish waters of the Baltic Sea, or the similarly "dwarfed" plants living on the tops of high mountains, or the thicker fur seen on individuals of a particular species living further north than other individuals of the same species, would not in some cases pass along those traits to their offspring, at least for several generations? In such a case I would call the form a variety. For me, varieties are defined by differences in closely related organisms that are passed to their descendants. Only such heritable differences in natural populations will concern us here.

The differences need not be large to merit our attention. Indeed, small differences among individuals in nature are of great importance. It seems unlikely that the sudden and considerable deviations of structure that we occasionally see in our domestic productions, particularly with plants, are ever permanently propagated in nature. Almost every part of every living being is so beautifully related to its complex conditions of life that it seems as improbable that any one part should have been suddenly produced as perfectly as we see it now as that a complex machine could be instantly invented by humans in a similarly perfected state.

Individual Differences

Now nobody supposes that all members of the same species are cast in exactly the same mold. Within every species we can see at least slight differences among individuals, even among the offspring from one set of parents. These individual differences are highly important for my argument, as they are often inherited and can thus provide the material for what I am calling "natural selection" to act on and accumulate, just as we can select for and accumulate—in any desired direction—individual differences in our domesticated animals and agricultural productions.

These individual differences generally affect what most naturalists consider to be an organism's unimportant parts. But I can provide a catalog of facts showing that an organism's "important" features, whether physiological or morphological, can also vary among individuals of the same species. Indeed, I am now convinced that even the most experienced naturalist would be surprised at how many cases of variability he could collect, even in functionally important structures, as I myself have done over many years of study. It should be remembered that taxonomists are not happy when they find variability in important characters: their focus is on finding *similarities* within groups, not differences. Remember, too, that not many people will laboriously examine and compare internal and other important organs in many individuals of the same species. If they did so, they would be surprised at what they would find. For example, I should never have expected that the branching of the main nerves close to the great central ganglion would vary as much as they do among individuals of a single insect species. Similarly, my friend and colleague, the well-respected entomologist Sir John Lubbock, has recently documented a fair degree of variability in these same main nerves among certain "scale insects," members of the insect genus *Coccus* (Figure 2.1), which may almost be compared

Figure 2.1 Scale insects (yellow), *Coccus viridis*.

to the irregular branching of the stem of a tree. Sir Lubbock has recently shown that the muscles also vary considerably among individuals in the larvae of certain insect species.

Some authors have argued that important organs never vary within a species, but when they make such claims they argue in a circle: these same authors, for practical purposes, rank only those characters that do *not* vary among individuals as important ones, while ignoring those characters that do vary. With such an approach, no instance of an important part varying will ever be found!

Questionable Species

Perhaps the most important organisms for my argument are those that seem in many respects to be distinct species, but which naturalists do not like to rank as distinct species because they so closely resemble some other forms or are so closely linked to other forms by organisms showing inter-mediate gradations. The boundary between varieties and species is indeed often uncertain, even as "species" and "varieties" are themselves difficult to define precisely, as mentioned earlier. In practice, when a naturalist can unite two forms together using other individuals that have characteristics intermediate between the two, he typically treats the one as a variety of the other, ranking the most common one as the species and the other as the variety; or sometimes he ranks the one that was described first as the spe-cies, and the other as the variety. But it is sometimes very difficult to decide whether or not to rank one form as a variety of another, even when they are closely connected by individuals with intermediate characteristics. Nor will the commonly assumed hybrid nature of the intermediate links always remove the difficulty. In very many cases, one form is ranked as a variety of another not because intermediate links are known, but rather because the observer supposes that such intermediate forms must exist somewhere, or may formerly have existed. And this opens a rather wide door for the entry of doubt and conjecture. Indeed, I refer to organisms which may or may not be true species as "doubtful" or "questionable" species.

In determining whether a form should be ranked as a species or as a variety, the opinion of naturalists having sound judgment and wide experience seems the only guide to follow. However, most well-marked and well-known varieties have been ranked as formal varieties by some workers, but as separate species by at least some other competent judges. Thus in such cases we must simply take the majority opinion, which is not fully convincing, or satisfying.

Varieties of such a questionable nature are quite common. Compare the several floras of Great Britain, France, or the United States, drawn up by different botanists, and see what a surprising number of forms have been ranked by one botanist as good species, and by another as mere varieties of a species. The esteemed botanist Mr. Hewett Cottrell Watson, to whom I am much obliged for assistance of all kinds, has marked for me 182 British plants that are generally considered to be varieties but which have all been ranked by some botanists as separate species. Even in making this list, though, he has omitted many trifling varieties that have been ranked by some botanists as separate species, and has entirely omitted some genera that are unusually polymorphic (i.e., highly variable in some characters within individual populations). Under genera, including the most polymorphic forms, the well-respected botanist Mr. Charles Cardale Babington discerns 251 species, whereas the equally respected botanist Mr. George Bentham gives only 112 species, a difference of 139 questionable forms!

Among animals that physically join for reproduction and which are highly mobile, forms that are ranked by one zoologist as a species and by another as a variety are rarely found within the same region, but are common in separated areas. It is incredible to see how many of those birds and insects of North America and Europe that differ only slightly from each other have been ranked by one eminent naturalist as undoubtedly distinct species and by another as varieties of one species, or, as they are often called, as "geographical races"! The esteemed naturalist Mr. Alfred Russel Wallace[1] has shown, in a series of valuable papers, that a variety of animals found among the different islands of the Malay Archipelago, especially members of the Lepidoptera (butterflies), can be placed into four categories: 1) variable forms, which show much variation among individuals living on individual islands; 2) local forms, which are quite uniform and distinct within each island, but differ somewhat from island to island, with the most widely separated forms being clearly distinct; 3) geographic races, which clearly differ from each other from island to island, but not so radically as to define them as separate species or merely as varieties; and 4) true representative species, which differ from each other in a number of

---

[1] Alfred Russel Wallace is the naturalist who independently came up with very similar ideas about evolution through natural selection, and in fact sent Darwin the draft of a paper on this topic in 1858, prompting Darwin to write his book on the origin of species.

clearly marked ways. However, nobody has come up with any set of criteria to clearly and objectively distinguish among variable forms, local forms, geographical races, and true species. In practice, the distinctions just aren't that easy to make.

Many years ago, I compared birds found on separate islands of the Galápagos Archipelago in the equatorial Pacific Ocean, both with each other and with birds found on the American mainland. Others have now made similar comparisons. Reviewing these comparisons, I was again much struck by how entirely vague and arbitrary the distinction was between species and varieties. Similarly, on the little Madeira group of islands, about 360 miles off the coast of North Africa, many of the insects characterized as varieties in Mr. Thomas Vernon Wollaston's admirable work would clearly be ranked as distinct species by many other entomologists. Even Ireland has some animals that are regarded as varieties by most zoologists but that have been ranked as separate species by others. Several very experienced ornithologists consider our British red grouse as only a strongly marked race of a Norwegian grouse species, whereas most ornithologists rank it as an undoubtedly distinct species peculiar to Great Britain. A large distance between the homes of two similar but distinct forms leads many naturalists to rank both as separate, distinct species…but exactly how large a distance is required to justify such a distinction? If that between America and Europe is sufficient, will the distance between the European continent and the Azores, or Madeira, or the Canaries, or Ireland, be sufficient as well?

We must admit that many forms considered by highly competent judges to be mere varieties have so perfectly the character of species that they are indeed ranked by other highly competent judges as good and true species. But to discuss whether they are rightly called species or varieties, before any definition of these terms has been generally accepted, is vainly to beat the air.

A good number of strongly marked varieties or questionable species are well worth thinking about more carefully, for several different and very interesting lines of argument have been used in attempting to determine their true rank.[2] Here I will give but one example, in considering the relationship between two flowering plants: the primrose (*Primula vulgaris*) (Figure 2.2A) and the cowslip (*Primula veris*) (Figure 2.2B). These plants are very different in appearance, smell differently and have different flavors, flower at different times, grow in different sorts of places, are found at different heights on mountains, and have different geographical ranges. Moreover, based on the results of the many experiments conducted by that most careful observer and plant hybridization expert Karl Friedrich von Gärtner, they can be hybridized, but only with difficulty. We could hardly wish for better evidence that the two forms are specifically distinct. On the other hand, they are united by

---

[2] Darwin omitted this paragraph from the Sixth Edition of *The Origin of Species*, but it seems worth including here.

(A)

(B)

Figure 2.2 (A) The primrose (*Primula vulgaris*). (B) The cowslip (*Primula veris*).

many intermediate links, and it seems unlikely that all of those links are simply hybrids. There is also what seems to me to be an overwhelming amount of experimental evidence showing that the primrose and cowslip have in fact descended from common parents, and consequently should be ranked as varieties of a single species. So the distinction between species and varieties is again not at all clear. Close investigation should eventually bring naturalists to agreement about how to rank such problematic organisms.

Intriguingly, it is where the organisms are best known that we find the greatest number of questionable species. I have been struck by the fact that if any animal or plant in nature be highly useful to people, or attract our attention for any other reason, varieties of it will almost universally be recorded. And at least some of these varieties will often be ranked by some authors as separate species. Look at the common oak, for example, a tree that has been much studied: while one particular German author divides the various forms of this plant into more than 12 distinct species, others generally consider them to be mere varieties of a single species. Similarly, in England the highest botanical authorities and practical men cannot agree whether the sessile (also called the Welch) oak and the pedunculated (also called the English) oak are good and distinct species in the genus *Quercus* or mere varieties of a single species. There is as yet no consensus.

When a young naturalist begins to study a group of organisms quite unknown to him, he is at first much perplexed to determine which differences to consider as species-specific traits and which ones as varieties, for he knows nothing of the amount or kinds of variation shown within the group. But if he confines his attention to one class of organisms within one particular geographical area, he will soon make up his mind how to rank most of the troublesome forms. Generally he will be initially impressed with the amount of difference in the forms that he is continually studying—just like the pigeon or poultry fanciers I talked about in Chapter 1—and thus will assign the various forms to many different species. When first starting off in this way, he has little general knowledge of analogous variation in other groups and in other geographical areas by which to correct his first impressions. However, as he extends the range of his observations, he will meet with more cases of difficulty, for he will come across a greater number of closely related forms. If his observations are extended widely enough, he will eventually be able to make up his own mind about which ones to call varieties and which to call species. But he will succeed in this at the expense of admitting that there is much variation among individuals—and the truth of this admission will often be disputed by other naturalists. Moreover, when he comes to study related forms brought here from countries that are well separated from each other—in which case he can hardly hope to find any intermediate links between his questionable forms—he will have to trust almost entirely to analogy, and his difficulties will then reach a climax.

Certainly no clear line of demarcation has yet been drawn between species and subspecies—that is, the forms which, in the opinion of some naturalists, are almost, but not quite, deserving the rank of separate species. Thus they are classified as subgroups within a particular species. Neither is there any clear objective distinction between the ranks of subspecies and well-marked varieties, or even between lesser varieties and simple differences among individuals. These differences all blend into each other in an insensible series, which impresses one's mind with the idea of an actual progression of forms.

Indeed, I look at such individual differences as being of great importance, though they may be of small interest to the taxonomist. Such small individual differences are in fact the first steps toward such slight varieties as are barely thought worth recording in works about natural history. But varieties that are in any degree distinct and permanent are, to me, early and important steps leading to more strongly marked and more permanent varieties; and I see these varieties as eventually becoming separate subspecies, and then separate species. In most cases, I attribute the gradual conversion of a variety—from a state in which it differs very slightly from its parent to one in which it differs more—to the action of natural selection in accumulating differences of structure in certain definite directions over long periods of time; I will explain this idea more fully later on. Thus, I believe that a well-marked variety is essentially an incipient species—a new species in the making. Whether or not you accept this idea will depend on the degree to which you are convinced by the facts and views given throughout this book.

We need not suppose that all varieties or incipient species will eventually become acknowledged as separate species. They may become extinct before that happens, for example, or they may stay as varieties for very long periods of time, as has been shown by Mr. Wollaston with the varieties of certain fossilized land snails in Madeira. If a variety were to flourish so greatly as to eventually become more numerous than the parent species, it would then be ranked as the species, and the original species would be ranked as the variety. Or the variety might eventually become so successful as to supplant and exterminate the parent species. Or both might coexist in comparable numbers and come to be ranked as two independent species. I will return to this matter later.

From these remarks, it will be seen that I look at the term "species" as one arbitrarily given for the sake of convenience to a set of individuals that resemble each other closely, and that it does not fundamentally differ from the term "variety," a term given to less distinct and more fluctuating forms.[3] The term "variety," again, in comparison with mere differences among individuals, is also applied arbitrarily, and merely for the sake of convenience. Variability is present wherever we look for it, even among individuals within any particular variety; it is just a matter of degree.

Wide Ranging, Much Diffused, and Common Species Vary the Most

Guided by theoretical considerations, I thought that some interesting results might be obtained regarding the nature and relationships of the species that vary the most by tabulating all the varieties in several well-documented plant groups. This turned out to be a much more complicated business than I had expected, but my good colleague Dr. Joseph Hooker has examined my tables and thinks that the statements I am about to make from them are fairly well established. These tables form the basis for most of the discussion that follows in this chapter.

The Swiss botanist Alphonse de Candolle and others have shown that plants which have very wide geographical ranges generally present us with many varieties; this might have been expected, as those plants are exposed to a wide range of physical conditions, and, even more importantly for my argument, they come into competition with many different sets of organisms in different parts of their range. But my tables further show that, in any particular region, the species that are the most common—that is, those that present the greatest number of individuals in the area—and the species whose members are the most widespread within their own country, often give rise to varieties sufficiently distinct to have been recorded in botanical works. Thus it is the most flourishing, that is to say the most dominant, species—those that range widely over the area, occupy the most diverse habitats in that area, and are the most numerous in individuals—which most often produce well-marked varieties, which as I have said, I consider to be incipient species. This, perhaps, might have been anticipated: in order for varieties to become permanent to any degree, they must have to struggle with the other inhabitants of the country, and win. Thus, species that are already dominant in an area will be the most likely to produce offspring which, although they may be in some slight degree modified, will still inherit those advantages that enabled their parents to dominate their compatriots in the first place.

---

[3] We now tend to define species by their reproductive isolation—i.e., their inability to mate successfully with members of other groups. In contrast, varieties of a single species can mate and successfully reproduce.

Species in Larger Genera Vary More Frequently Than Those in Smaller Genera

Now some genera contain many species—I will call these the "larger genera"—while other genera (the "smaller genera") contain many fewer species (see Figure 1.5). If the plants inhabiting any particular country and described in any botanical manual are divided into two groups, with all those in the larger genera (containing many species) being placed on one side of the page and all those in the smaller genera (containing fewer species) being placed on the other side of the page, a somewhat larger number of the most common and more widespread species will be found on the side listing the larger genera. That is, the genera containing the most species tend to also contain the most common and widespread species. Again, this might have been expected, for the mere fact that whenever we see many species of the same genus inhabiting any one country, there must be something in the conditions of that country—biological or physical, or both—that are especially favorable to members of that genus. Consequently, we would expect to find a larger proportion of dominant species in the larger genera...and indeed we do. But actually I am surprised at seeing so clear a finding, as there are many factors acting to obscure such as result. Here I will mention only two of these factors. First, freshwater plants and salt-loving plants (e.g., salt-marsh plants) generally have very wide ranges and are much diffused, but this seems only to reflect the sorts of specialized habitats that they occupy rather than having anything to do with the size of the genera to which the species belong. Secondly, plants that are low on the scale of organization (mosses, for example) are generally much more widely diffused than plants higher on the scale; and here again there is no close relation to the number of species within the genera. And yet even with these sorts of complicating factors, my tables still show that at least a small majority of the most dispersed and dominant species belong to the larger genera.

In looking at species as being essentially strongly marked and well-defined varieties, I logically expected that the species of the larger genera in each country would more often present varieties than the species found in the smaller genera; for wherever many closely related species (i.e., those in the same genus) have been formed, many varieties or incipient species ought, as a general rule, to be now forming. Where many large trees grow, we expect to find saplings. Where many species of a particular genus have been formed through variation, circumstances there have clearly favored variation. On the other hand, if each species was formed through a special act of creation, there is no apparent reason why more varieties should occur in a group having many species than in one having only a few.

As a way of testing my expectation, I have arranged the plants of 12 countries, and the beetles (members of the order Coleoptera) of two districts, into two nearly equal groups, the species of the larger genera again on one side and those of the smaller genera on the other side. It indeed turns out that a larger proportion of the species on the side of the larger genera present varieties than do those on the side of the smaller genera.

Moreover, the species belonging to the larger genera invariably present a larger *average* number of varieties than do the species belonging to the smaller genera. Both these results follow even when all the smallest genera (each containing only between one and four species) are completely excluded from the tables.

Clearly, species are little more than strongly marked and long-lasting varieties. For wherever many species of the same genus have been formed, or where, if we may use the expression, the "species-manufacturing apparatus has been active," we ought generally to find that factory still in action, particularly as we have every reason to believe that the process of manufacturing new species is a slow one. And this certainly is the case, if we think of varieties as incipient species: my tables clearly show the general rule that wherever many species of a genus have been formed, the species of that genus present an exceptional number of varieties; that is, of incipient species beyond the average. It is not that all large genera are now varying much and are thus increasing in the number of their species, or that no small genera are now varying and increasing; for if this had been so it would have been fatal to my theory. Geology plainly tells us that small genera have often increased greatly in size over long periods of time, and that large genera have often reached a maximum number of species, declined, and then disappeared. **My point here is simply that where many species of a genus have been formed, on average many new species are still forming.**

Many Species Included in the Larger Genera Resemble Varieties: They Are Closely Related, and Have Restricted Ranges

Several other relationships between the species belonging to large genera and their recorded varieties deserve our attention. We have seen that there are no infallible criteria by which to distinguish species from distinct varieties; in those cases in which intermediate links have not been found between questionable forms, naturalists must reach a determination based on the amount of difference between them, judging by analogy whether or not that amount is sufficient to raise one or both to the formal rank of species. Thus the amount of difference is one very important criterion in settling whether two forms should be ranked as species or as varieties. Now the Swedish botanist and mycologist Elias Fries has remarked with regard to plants from genera containing many species (i.e., in large genera), that the amount of difference between the species is often exceedingly small. The English entomologist John Obadiah Westwood has made the same point with regard to insects. I have tried to test this through some calculations, and as far as my imperfect results go, they confirm those views: in genera containing an especially large number of species, the amount of difference between the species is indeed quite small. I have also consulted some sagacious and experienced observers, and, after careful thought, they all agree with this view. In this respect, therefore, the species of the larger genera resemble varieties more than do the species of the smaller genera. Said another way, in the larger genera, in which a number of varieties or incipient species greater than the average are now in the process of being manufactured, many of the species already manufactured still to a certain extent resemble varieties, for they differ from each other by a less than usual amount of difference.

In addition, the species found among the larger genera are related to each other in the same manner as the varieties of any one species are related to each other. No naturalist pretends that all the species of a genus are equally distinct from each other; they may generally be divided into sub-genera, or sections, or even lesser groups. As Mr. Fries has remarked, little groups of species are generally clustered like satellites around certain other species. And what are varieties but groups of forms, unequally related to each other and clustered around certain other forms—that is, around their parent species?

Undoubtedly there is one most important point of difference between varieties and species: the amount of difference between varieties, when compared with each other or with their parent species, is much less than that between the species of the same genus. But when we come to discuss the principle that I am calling "Divergence of Character" in Chapter 4, we shall see how this may be explained, and how the relatively smaller differences between varieties will tend to increase over time into the greater differences between species.

There is one other point that I think deserves mention. Varieties generally have much restricted ranges: indeed, if a variety were found to have a wider range than that of its supposed parent species, their designations should be reversed, and the species be called the variety and the variety be called the species. But it seems that species that are very closely allied with other species, so that the different species essentially resemble varieties of each other, also often have much restricted ranges. For instance, the English botanist Mr. Watson, mentioned earlier, has marked for me 63 plants that are ranked as species in the well-sifted *The London Catalogue of British Plants* (4th edition), but which he considers as so closely related to other species that their status as separate species is questionable; these 63 reputed species range on average over 6.9 of the provinces into which Mr. Watson has divided Great Britain. Now, in this same catalog, 53 acknowledged *varieties* are recorded, and these range over 7.7 provinces, whereas the species to which these varieties belong range over 14.3 provinces. Thus, the 53 acknowledged varieties have very nearly the same restricted average range as have those very closely related forms marked for me by Mr. Watson as questionable species, but which are almost universally ranked by other British botanists as good and true species.

Summary

In summary, **varieties have the same general character as species**, for they cannot be distinguished from species except by first discovering intermediate linking forms and second by showing a certain amount of difference, which is largely undefined: when two forms differ very little from each other they are ranked as varieties, even when intermediate forms linking the two have not been discovered. But the amount of difference considered necessary to rank two forms as species is quite unspecified. In genera having more than the average number of species in any country, the species within these large genera have more than the average number of varieties. In large genera, the species are apt to be closely, although unequally, allied together, forming little clusters around certain species. Species very closely allied to other species apparently have restricted ranges. In all these several respects the species of large genera present a strong analogy with varieties within species. **We can clearly understand these analogies if we accept that what are now distinct species once existed as mere varieties**, and thus originated as varieties; on the other hand, these well-documented analogies would be utterly inexplicable if each species had been independently created.

We have also seen that it is the most flourishing or dominant species of the genera containing the most species (the "larger" genera in my terminology) that on average vary the most, and it is these "varieties," as we shall see in more detail later, that tend to become converted into new and distinct species. The larger genera thus tend to become larger over time, and throughout all of nature the forms of life that are now dominant tend to become still more dominant over time, by leaving many modified and dominant descendants. But through steps to be explained later in this book, the larger genera also tend to eventually break up into small genera. And thus the forms of life throughout the world slowly become divided into groups within groups (see Figure 1.5).

Key Issues to Talk and Write About

1. Find out two interesting things about one of the people that Darwin mentions in this chapter. Choose from the following: Elias Fries, Hewett Cottrell Watson, Joseph Hooker, Alphonse de Candolle, or Alfred Russel Wallace.

2. Read the paragraph (page 36) that begins "Intriguingly, it is where the organisms are best known that we find...." Now, write a one-sentence summary of that paragraph: How would Darwin make his key points if he only had one sentence in which to do so? First list the two to three major points that Darwin is making. Then try to get all of those points into a single sentence. The sentence must be accurate, contain each of the major points, and be in your own words,

as though you were explaining something to a friend. It should also make sense to a reader who has never read the original paragraph.

3. How does Darwin define "large genera" (see the section on page 35 that begins, "Now some genera contain many species—I will call these the 'larger genera'")? How does Darwin's example about the amount of variability seen among species within large genera fit in with his argument that what we now see as distinct species were once merely varieties of some other, ancestral species?

4. Here is a sentence from the original version of Darwin's Chapter 2:

   "That varieties of this doubtful nature are far from uncommon cannot be disputed." Try rewriting that sentence to make it clearer.

5. We now generally define species as being reproductively isolated units, whereas varieties within a species (the various breeds of dogs, for example) are able to mate successfully with each other and produce viable offspring. How does Darwin explain the difficulty of distinguishing between varieties and species based on physical characteristics?

Online Resources *available at* sites.sinauer.com/readabledarwin

Links

2.1 Interesting information about primroses and other plants with dimorphic or trimorphic flowers, and why Darwin was so interested in them
http://darwinsflowers.wordpress.com/exhibit/function-of-flower-forms/

2.2 A brief summary of the difficulties in defining species
www.evolution.berkeley.edu/evosite/evo101/VADefiningSpecies.shtml

(Note: Web addresses may change. Go to sites.sinauer.com/readabledarwin for up-to-date links.)

Bibliography

Watson, H. C., ed. 1859. *The London Catalogue of British Plants* (4th edition). London.

3

The Struggle for Existence

In this chapter, Darwin emphasizes that for all plants and all animals in nature, far more individuals are born than can possibly survive; predation, competition, and physical stresses all take their toll, often in amazingly complex and surprising ways. Competition will be most intense among individuals that are the most similar to each other (i.e., among members of the same species). This competition will, in a sense, have individuals pushing each other over many generations—through survival, reproductive success, and the inheritance of favorable characteristics—to become better and better adapted to their way of life.

Let me set the stage: How does the "struggle for existence" relate to what I am calling "natural selection"?

I have already shown in Chapter 2 that in the natural world, individuals within any group vary in a great many traits. Indeed, this is something that everyone seems to agree with. But the mere existence of this variation among individuals, although it provides the foundation for all my thinking, doesn't explain how new species arise in nature, or how individual organisms have become so perfectly adapted to their surroundings and lifestyles. We see such beautiful adaptation very clearly in the woodpecker (Figure 3.1) and in the mistletoe, for example, and in the simplest parasite that clings to the hairs of a dog or a sheep, or to the feathers of a bird, and we see the perfect structure of beetles that dive down through the water to feed, and in the feathered plant seeds (Figure 3.2) that are wafted away by even the gentlest breeze. Basically we see beautiful, marvelous adaptations everywhere we look, in every part of the living world.

How have all those exquisite adaptations come about? And how can small variations in various traits eventually give rise to new species? How can varieties, which I have called "incipient species" (i.e., species in the making), eventually become converted into distinct, separate species—i.e., species that generally differ from each other more than do varieties of the same species?

Figure 3.1 The red-bellied woodpecker, male (*Melanerpes carolinus*).

Figure 3.2 Seeds of the common dandelion (*Taraxacum officinale*) being dispersed away from the parent plant.

It all follows from the *struggle for life*, a struggle most eloquently discussed in the recent writings of the eminent geologist Sir Charles Lyell and the Swiss botanist Augustin Pyramus de Candolle. Because of this struggle, even the smallest of variations can help improve the chances of an individual's survival. **Offspring that then inherit those particularly advantageous traits from their parents will have a better chance of surviving than those that do not, because—and here is one of my main points—from the great many individuals of any species that are born in any one year, only a small number can survive.** I call this basic principle "natural selection," to help us see its connection to our own great powers of selection in breeding animals and plants for our own use (see Chapter 1). But perhaps Mr. Herbert Spencer's term "survival of the fittest" is more accurate.[1]

In Chapter 1, I gave examples of how we have molded domesticated animals and plants for our own use, selecting for the gradual accumulation of slight but useful variations by carefully choosing which individuals to breed with other individuals in each of many generations, taking advantage of the natural variation among individuals given to us by the hand of nature. But what I have now called *natural* selection is as immensely superior to our feeble efforts at selection as the works of nature are to those of art, as you will see more clearly as you read further in this book.

Everything about the struggle for existence follows from this simple fact, as shown by the Swiss botanist Alphonse de Candolle (the son of Augustin Pyramus de Candolle who was mentioned earlier) and Sir Lyell: **All organisms are exposed to severe competition, predation, and physical stress. We must keep this constantly in mind: If we don't integrate this idea into our brains—if we don't keep this continuous struggle in mind at all times—the**

---

[1] This term is commonly misunderstood: yes, natural selection is about survival, but it is especially about fitness for reproductive success, which includes survival.

whole point of our observations on animal and plant distributions, rarity, abundance, extinction, and variation will only be dimly seen, or will be completely misunderstood.

When we look on the face of nature, seemingly bright with gladness, we often see what appears to be a superabundance of food. What we don't think about is that the birds that are merrily singing all around us must in fact eat insects and seeds to survive, and thus are constantly destroying other forms of life. And we forget how these songsters, or their eggs and babies, are themselves destroyed by predators, including other birds. And we often forget that although food supplies may indeed be plentiful at some times of the year, that is certainly not true in all seasons and in every year; and at those times, individuals must indeed compete and struggle with each other for their continued existence.

The Term "Struggle for Existence" Used in a Larger Sense

When I talk about the "struggle for existence," I'm not just talking about how different organisms depend on each other, and I'm not just talking about the struggle to stay alive; I'm also talking about success—or not—in leaving offspring that grow up to reproduce successfully themselves. Two wild dogs, for example, may fight over food when food is in short supply, and that fight may determine which dog eats and which does not, and in fact may determine which one lives to reproduce.

But a plant on the edge of a desert also struggles for life, this time against dryness—for it depends on moisture for its existence. A plant may produce 1,000 seeds every year, but if only one of those 1,000 seeds survives to re-productive maturity, then that one surviving plant has truly struggled with other plants, both of its own species and of other species, for its survival. Similarly, mistletoe seedlings growing on a single branch compete with each other for space on that branch. But even more interesting is that mistletoe is moved about and transported to other areas by birds, so that the contin-ued existence of mistletoe depends on its being spread by those feathered chariots. Thus we can say that the mistletoe, in this sense, also "struggles" with other fruit-bearing plants, in trying to get birds to eat only them, thus distributing its own seeds instead of those produced by other plants.

The term "struggle for existence" applies in all of these instances.

Exponential Rates of Increase

A struggle for existence in the natural world is inevitable, because of the high rate at which most organisms reproduce themselves. Every animal species, whether its members produce several eggs or many eggs each season, must suffer great destruction during some phase of its life, and during some sea-sons, at least in some years. The same must be true for plants, whether they produce several seeds, or many seeds. Otherwise, based on the principle

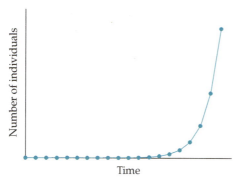

Figure 3.3 An example of exponential growth. The population increases over time in proportion to the number of individuals already present: the more individuals present at any given time, the faster the population grows.

of exponential increase, their numbers would quickly become so enormously great that no one country could support them all (Figure 3.3)! By "the principle of exponential increase" I mean that population growth rates increase proportionally as the size of the population increases, as first explained by Thomas Malthus in 1798. For example, if we have ten females in a population and each female produces two eggs, the population will grow by another 20 individuals in the next generation if all of the eggs hatch and no individuals die. But if we have 1,000 females in the population, that population will grow by 2,000 individuals in the next generation under the same rules. When we have 10,000 females in the population, that population will grow by 20,000 individuals in the same amount of time, again assuming no mortality: the increase in numbers is proportional to the size of the population.

A struggle for existence is inevitable whenever more individuals are produced than can possibly survive. Individuals struggle with others of the same species, or with the individuals of some other species, or with the physical stresses of life, such as temperature changes or dehydration; but struggle they must. Here, the rule of Mr. Malthus applies with tremendous force to all living organisms in the natural world, both animals and plants; in this situation there can be no artificial increase in food supplies, and no means of keeping individuals from mating and reproducing. Although some species may well be increasing in numbers now, more or less rapidly, all cannot do so, and none can do so forever, for the world would not hold them.

There is no exception to the rule that every species naturally increases at so high a rate that, without high rates of mortality, the earth would soon be covered by the offspring of a single pair of parents. Even humans—which breed fairly slowly compared with most other animals—have doubled their population numbers in just the past 25 years; at this rate, in the absence of mortality, within a thousand years there would be no standing room for our descendants! Indeed, the Swedish botanist and zoologist Carl Linnaeus calculated (in the 1730s) that if an annual plant produced only two seeds in its lifetime—something that no plant actually does—and if their seedlings produced two seeds themselves the following year, and so on each year, without any mortality there would be...one million plants in only 20 years!

Here is another example. The elephant is believed to be the slowest breeder of all known animals. Let's assume that it doesn't start reproducing until it is 30 years old, and that it then continues to reproduce every year for the next 60 years, leaving six offspring each year. Suppose, too, that the animals die when they become 100 years old. According to my calculations, then, after about 740 or 750 years, there would be almost 19 million elephants alive on this planet, all descended from that first pair.

But this is not just theorizing on paper. In fact, we have many actual examples of astonishingly rapid increases in the numbers of some animals in the real world, when living under favorable conditions for even just two or three seasons. Even more striking is the evidence from many of our domesticated animals that have escaped from domestication and lived subsequently in the wild in some parts of the world; indeed, if the reports about the remarkably high rates of increase of slow-breeding cattle and horses in South America, and more recently in Australia, had not been very well documented, they would have been too incredible to be believed!

The same is true with plants: I know of many cases in which introduced plant species have become common throughout entire islands in less than ten years. Indeed, several such plants, including the cardoon (*Cynara cardunculus*) (Figure 3.4) and a species of tall thistle, were introduced from Europe to the plains of La Plata in Argentina only a few years ago and are now the commonest plants in that new habitat; they now dominate many, many square miles of surface and in fact pretty much exclude all other plants. Similarly, some plants introduced from America sometime after its discovery now range in India over enormous distances, from Cape Comorin to the Himalaya Mountains. How can we explain this? Certainly in these and in many other such examples that I could give, it's not that the egg production of the introduced animals or plants has suddenly and temporarily increased to any great extent. No, the obvious explanation is that the conditions of life in the new lands have been extremely favorable to the invaders: so favorable, in fact, that both old and young have survived very well, and most of the young have managed to reproduce successfully. Their exponential rate of increase in their new homes, the result of which is always remarkable and surprising, completely explains their extraordinarily rapid increase in numbers and the tremendous spread of their populations in their new homes.

In the natural world pretty much every full-grown plant produces seed every year. Among

Figure 3.4 The cardoon (*Cynara cardunculus*).

Figure 3.5 The fulmar petrel (*Fulmarus glacialis*).

animals, too, there are very few species that don't mate and reproduce each year. Thus we can state with confidence that all plants and animals tend to increase at an exponential rate—and that all species would rapidly dominate every place that supported their existence—unless the exponential tendency to increase was somehow suppressed by destruction at some point in the life cycle.

Cows, sheep, and the other large domestic animals that we are all so familiar with are not good examples of life in the wild at all, because we see no great destruction falling on them. But in fact we kill thousands of those animals every year for food; an equal number would be killed in one way or another in the natural world.

Note, too, that it generally matters little whether an organism produces thousands of eggs or seeds each year, or only a very few. In both cases the offspring would eventually dominate the landscape; one would take a few more years to do so than the other, but eventually both would reach huge population sizes. The condor lays only a few eggs each year, while the ostrich lays a dozen, and yet in some countries the condor population is larger than that of the ostrich. Why, the fulmar petrel (*Fulmarus glacialis*) (Figure 3.5) lays only one egg each year and yet may be the most numerous bird in the world!

Producing many eggs *is* of some importance, of course, to species whose food supply fluctuates from year to year, for it allows them to rapidly expand their population size when conditions are good. But the main benefit of producing a great many eggs or seeds is to make up for the great destruction that occurs at some point in the life cycle, usually at some early point. If many eggs or young are routinely killed by natural forces, then many must also be produced, or the species will become extinct. You could have no change at all in the number of trees in an area for a species that lived an average of 1,000 years, even if it produced only one seed in its lifetime, as long as all those seeds germinated successfully and in suitable locations. Thus, in every case we can think of, the average number of animals or plants now living in an area depends very little and only indirectly on the number of eggs or seeds it produces.

Never forget that every living thing may be trying its utmost to increase in numbers, that each lives by struggle at some point in its life, and that in every generation, sometimes at repeated intervals, all populations experience heavy destruction. Reduce that level of natural destruction even slightly, and the number of individuals living in an area will increase greatly in a very short time.

Nature of the Checks to Population Growth Increase

The factors acting against the natural tendency of each species to increase in numbers are most obscure. Look at the most vigorous species: by as much as it swarms in numbers, so much will its numbers tend to increase still further in the next generation. What, then, limits its growth? Unfortunately, we can't identify all the various checks to population growth, even in a single instance.

I hope to discuss the topic more fully in the future, but here I will just bring up some of the chief points. Firstly, it seems that for animals, the eggs or the very young usually suffer the most. Similarly, with plants we see a vast destruction of seeds. My observations suggest that seedlings have the most difficulty germinating in areas that are already thickly stocked with other plants. Seedlings are also destroyed in vast numbers by various enemies. For example, I cleared an area of ground 3 feet by 2 feet in my garden, so that there could be no choking from other plants, and I then separately marked each of the seedlings of our native weeds as they germinated. Out of 357 such seedlings that sprouted, 295 (nearly 83%) were destroyed, chiefly by slugs and insects.

When any area of lawn that has long been mowed or closely grazed by cattle is let to grow, the more vigorous plants gradually kill the less vigorous plants, even if those other plants are full grown. Thus, out of 20 species of plants growing on a little 3-foot by 4-foot plot of turf at my home, 9 species perished simply from other species being allowed to grow up freely around them.

The amount of food available must of course set the upper limit to population size in any area, for any animal species. But often it is not the availability of food that determines population size for a species, but rather the extent of *predation* by other animals. Thus, there seems little doubt that the population of partridges, grouse, and hares on any large estate depends chiefly on their destruction by predatory foxes, weasels, and other vermin. Indeed, if not a single game animal were shot during the next 20 years in England, and at the same time none of these predators were destroyed, there would, in all probability, be fewer game animals than at present even though hundreds of thousands of such game animals are now killed every year. On the other hand, as in the case of the elephant and rhinoceros, for example, sometimes predation is not so important in regulating population growth; even the tiger in India very rarely dares to attack a young elephant protected by its mother. The importance of predation in regulating population size clearly varies with species and situation.

Climate also plays an important role in determining species numbers; indeed, I believe that long periods of extreme cold or drought are probably the most effective of all checks on population growth. I estimated that the winter of 1854–1855 destroyed 80 percent of the birds on my own property; this is a tremendous amount of destruction when we remember that epidemics in humans rarely kill more than 10 percent of the individuals in a population.

At first glance, the action of climate might seem largely unrelated to the struggle for existence that we have been discussing. However, to the extent that climate change acts chiefly to reduce food availability, it can bring on the most severe struggle between individuals, both of the same species and of different species, that subsist on the same kinds of food. Even when climate—periods of extreme cold, for example—acts on organisms directly, it will be the least healthy individuals, or those that have gotten the least food through the advancing winter, who will suffer the most.

When we travel from north to south, or from a damp region of the country to a dry one, we invariably see some species gradually getting rarer and rarer, and finally disappearing altogether; the change of climate being so conspicuous, we are tempted to attribute the shift in species numbers entirely to its effect. But this is a false view: we forget that each species, even where it is most abundant, is constantly suffering enormous destruction at some period in its life, either from enemies or from competitors for its space or food. If those enemies or competitors are in the least degree favored by any slight change in climate, they will increase in numbers and, as each area is already fully stocked with inhabitants, the other species will decrease in abundance. When we see a species decreasing in numbers as we travel southward, we may feel sure that the cause lies quite as much in other species being favored as in this species being hurt by the changing climate.

We see the same thing when traveling from south to north, although to a somewhat lesser degree, for the numbers of species of all kinds, and therefore the numbers of competitors, decreases as we travel northward. Thus in going northward, or in ascending a mountain, we meet with stunted forms, due to the directly injurious action of climate, more often than we do in proceeding southward or in descending from a mountain. When we reach the Arctic regions, or snow-capped summits, or absolute deserts, the struggle for life is almost exclusively against the physical environment rather than with other species.

However, climate often acts largely indirectly, by favoring other species: We may clearly see this in the prodigious numbers of plants in our gardens that, although they have no difficulty coping with our climate, nevertheless never become "naturalized," (i.e., part of the natural landscape in their new homes); they just cannot compete with our native plants, nor can they avoid being destroyed by our native animals.

When a species, owing to highly favorable circumstances, shows an inordinate increase in numbers in a small area, epidemics often ensue; certainly we see this often with our game animals. And here we have a limiting check on population growth that seems to be independent of the struggle for life. But even some of these so-called epidemics appear to be caused by parasitic worms, which have for some reason, possibly in part though the ease of spread amongst the crowded animals, been disproportionately favored: and thus we see here another sort of struggle, this one between the parasite and its host.

For many species, population size must be very large relative to the number of its enemies if the species is to be maintained. Thus, for example, we can easily raise plenty of corn and rapeseed in our field, because there are so many seeds compared with the number of birds that feed on them. Nor can the birds, though having a superabundance of food at this one season, increase in number proportionally to the supply of seed, as their numbers are held back in the winter, when food is scarce. But anyone who has tried knows how troublesome it is to get seed from only a few wheat plants or other such plants in a garden; I myself have lost every seed in such cases.

This idea of a large population size being essential to the preservation of a given species explains, I believe, some singular facts in nature—for example, that some very rare plants can sometimes be extremely abundant in the few spots where they do occur, and that some plants can be extremely abundant even on the extreme edges of their ranges. In such cases it seems that a particular plant can exist only where the conditions of its life were so favorable that many can exist together, and thus save each other from utter destruction. I should add that the good effects of frequent intercrossing and the ill effects of mating with close relatives probably come into play in some of these cases; but I will say no more about this intricate subject here.

Complex Relations of All Animals and Plants to Each Other in the Struggle for Existence

We have many records showing how complex and often unexpected the checks and relationships are between organisms that have to struggle with each other in the same area. I will give only one example, one which has interested me a great deal despite its simplicity. One of my relatives in Staffordshire, England, allowed me to explore his estate in detail. On that estate there was a large and extremely barren heath (Figure 3.6), one that had never been touched by the hand of man. But on the same estate there were several hundred acres of exactly the same sort that had been enclosed

Figure 3.6 Barren heathland in Dartmoor National Park, U.K.

by fencing 25 years earlier, and planted with Scotch fir trees. The change in the native vegetation of the planted part of the heath was most remarkable, and more than is generally seen in passing from one quite different sort of soil to another. Not only were the proportional numbers of the various heath plants wholly different in the two areas, but 12 plant species (not even counting the grasses and grass-like sedges) that flourished in the plantations could not be found at all on the heath. The effect on the insect populations must have been even greater, for six species of insect-eating birds that were very common on the plantations were not seen at all on the heath. And the heath itself was visited by two or three distinct insectivorous bird species that were not seen on the plantations. Thus we see how potent the effect can be, of simply introducing a single new tree species, nothing else having been done other than to enclose the land so that cattle could not enter.

But in an area near Farnham, in Surrey, England, I plainly saw just how important enclosure itself can be. Here there are extensive heaths, with a few clumps of old Scotch firs on the distant hilltops. Within the last ten years large spaces on those heaths have been enclosed, and self-sown fir trees are now springing up in multitudes, growing so close together in fact that all cannot live. After I learned that these young trees had not been deliberately sown or planted within the enclosures, I was so surprised at their great numbers that I looked at them from several different vantage points, taking in hundreds of acres of unenclosed heath in each view; literally, I couldn't see a single Scotch fir in any of the unenclosed areas, except for the old planted clumps. But on looking closely between the stems of the heath in these unenclosed areas, I found a multitude of seedlings and little trees that had been perpetually grazed down by cattle. In one square yard, at a point some hundred yards distant from one of the old Scotch fir clumps, I counted 32 little trees. One of them, with 26 growth rings on its stem, had during many years tried to raise its head above the other stems of the heath, and had failed to do so. No wonder then, that as soon as some portion of the land was enclosed, keeping out all cattle, it became thickly clothed with vigorously growing young firs. Yet the unenclosed heath was so extremely barren and so extensive, that no one would ever have imagined that cattle could have so closely and effectively searched it for food.

Here then we see that cattle absolutely determine the existence of the Scotch fir in this part of England. But in some parts of the world, insects determine the existence of the cattle. Perhaps Paraguay, near the center of South America, offers the most curious example. Neither cattle nor horses nor dogs have ever run wild in Paraguay, although they certainly swarm southward and northward in a feral, undomesticated state. Two naturalists, Félix de Azara of Spain and Dr. Johann Rudolf Rengger of Switzerland, have shown that this is caused by large populations of a certain fly in Paraguay: the fly lays its eggs in the navels of cattle and other mammals. Population growth of these flies, numerous as they are, must be habitually checked by some agent, probably birds. Thus, if insect-eating birds (whose numbers are, in turn, probably regulated by hawks or other beasts of prey) were to increase in Paraguay, the fly population would decline. Cattle and horses could then become wild, and that would greatly alter the vegetation of the region, as indeed I have observed elsewhere in South America. This again would largely affect the insects, and thus affect the insectivorous birds, just as we have seen in Staffordshire; and so onward in ever-increasing circles of complexity.

We began this series with insectivorous birds, and we have ended with them. Of course, in nature the relationships can never be this simple. Battle within battle must ever be recurring with varying success, and yet in the long run the forces are so nicely balanced that the general abundances and distributions of organisms often remain uniform for long periods of time, though assuredly the merest trifle would often give the victory to one species over another. Even so, so profound is our ignorance, and so great our presumption, that we marvel when we hear about the extinction of any organism. And as we do not see or understand the cause of that extinction, we invoke biblical floods and other cataclysms that desolate the world, or invent laws about predetermined, "fixed lifespans" of species.[2] But no: the causes in fact all relate to the struggle for existence.[3]

Allow me to give just one more example showing how plants and animals, so different from each other in the scale of nature, are bound together by a web of complex relations. Later in this book I will show that the exotic plant *Lobelia fulgens*, the "cardinal flower" (Figure 3.7A), is never visited by insects in this part of England

Figure 3.7 (A) Cardinal flower (*Lobelia fulgens* = *L. cardinalis*). (B) Heartsease (*Viola tricolor*). (C) Red clover (*Trifolium pratense*).

---

[2] In trying to explain extinctions, some authors had suggested that species are created with fixed lifespans— i.e., that they are essentially programmed for automatic extinction after a certain number of years.

[3] While that has indeed generally been the case, five major extinctions are now known to have been caused by large-scale cataclysms, such as that occurring about 66 million years ago, which killed off all of the non-feathered dinosaurs and about 75 percent of all other animal and plant species on Earth.

and consequently, because of its peculiar structure, it never reproduces successfully here. Similarly, many of our orchids absolutely require moths to visit and remove their pollen masses, and thus to fertilize them. I also have reason to believe that bumblebees are indispensable to the fertilization of the heartsease (*Viola tricolor*) (Figure 3.7B), for I have seen no other types of bee visit this flower. From experiments that I have recently conducted at my home, I have found that some kinds of clover are fertilized only by bees. But only bumblebees visit the red clover (*Trifolium pratense*) (Figure 3.7C), as other bees cannot reach the nectar within the long, tubular flowers of this species. Therefore I have little doubt that if the entire genus of bumblebees became extinct, or even just very rare, in England, the heartsease and red clover would also become very rare, or perhaps disappear entirely.

Now the number of bumblebees in any district depends to a great degree on the number of field mice living there, because the mice destroy the bee's honeycombs and nests. Indeed, Mr. Henry Wenman Newman, who has studied the habits of bumblebees for many years, believes that "more than two-thirds of them are thus destroyed all over England." As everyone knows, the number of mice largely depends on the number of cats: as Mr. Newman says, "Near villages and small towns I have found the nests of bumble-bees more numerous than elsewhere, which I attribute to the number of cats that destroy the mice." Thus it is quite likely that the presence of large numbers of cats in a district might determine, through the intervention first of mice and then of bees, the commonness of certain flowers in that district—remarkable!

In the case of every species, many different checks probably come into play, each acting at different times of life and during different seasons or years; perhaps one check or some few are especially potent, but all ultimately combine in determining the average number of individuals in an area, or even the existence of the species. In some cases it can be shown that very different checks act on the same species in different areas.

When we look at the diversity of plants and bushes clothing an entangled bank along a river, we are tempted to attribute their proportional numbers and kinds to mere "chance." But how false a view this is! Everyone has heard that when an American forest is cut down, a very different vegetation springs up. But it has also been observed that ancient Indian ruins in the Southern United States, which must first have been cleared of trees many years ago, now display the same beautiful diversity and kinds of trees in the same proportion as in the surrounding virgin forests; the forest seems to have returned now to its former state. What a struggle between the several kinds of trees must have gone on here during long centuries, each tree annually scattering its seeds by the thousands; what war between insect and insect—and between insects, snails, and other animals with birds and beasts of prey—all striving to live and reproduce, and all feeding on each other or on the trees, or on the seeds and seedlings of those trees, or on the other plants that first clothed the ground and thus checked the growth of the trees! Throw up a handful of feathers, and all must fall to the ground according to definite laws; but how simple is that problem compared with trying to understand the actions and reactions of the innumerable plants and animals that have determined, over the course of centuries, the proportional numbers and kinds of trees now growing on the old Indian ruins!

The dependency of one organism on another, as with a parasite dependent on its host, lies generally between beings very different from each other in the scale of nature. This is also often the case with those that may strictly be said to struggle with each other for existence, as in the case of locusts and grass-feeding quadrupeds, such as cattle. **But almost invariably the struggle will be most severe between individuals of the same species, for they frequent the same districts, require the same foods, and are exposed to the same dangers.** In the case of varieties of the same species, the struggle will generally be almost equally severe, and we sometimes see the contest quickly decided. For instance, if several varieties of wheat are sown together, and the mixed seed is re-sown for the next generation, those varieties that best suit that particular soil or climate, or are naturally the most fertile, will beat the others and so yield more seed; consequently, in a few years, these varieties will quite supplant the other varieties. It takes a good deal of work to keep up a mixed stock of even such extremely close varieties as the variously colored sweet peas: each variety must be harvested separately each year, and the seed then mixed in due proportion; otherwise the weaker kinds will steadily decrease in numbers and eventually disappear.

And so it is again with the varieties of sheep: certain mountain varieties will apparently starve out other mountain varieties, so that they cannot be kept together. The same result has followed from keeping together different varieties of the medicinal leech: one variety eventually wins out over the others. It may even be doubted whether the varieties of any one of our domesticated plants or animals have so exactly the same strength, habits, and constitution, that the original proportions of a mixed stock could be maintained for even a half-dozen generations, if they were allowed to struggle together like beings in a state of nature, and if the seed or the young were not deliberately sorted by us every year.

The Struggle for Life Is Generally Most Severe between Individuals and Varieties of the Same Species

This idea, that the struggle for life is more severe the more closely the competing individuals resemble each other, leads in many interesting directions. **Because species within the same genus are usually similar in habits and constitution, and always have similarities in structure, the struggle will generally be more severe between species of the same genus when they come into competition with each other, than between species of different**

(A)   (B)

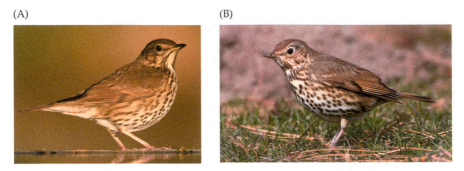

Figure 3.8 (A) Mistle thrush (*Turdus viscivorus*). (B) Song thrush (*Turdus philomelos*).

genera. For example, the recent extension over parts of the United States of one particular swallow species has caused the population of another swallow species to decrease. Similarly, the recent increase of the mistle thrush (Figure 3.8A) in parts of Scotland has caused populations of the song thrush (Figure 3.8B) to decrease. And how frequently we hear of one species of rat taking the place of another species under the most different climates! In Russia, the small Asiatic cockroach has driven out everywhere before it a larger species in the same genus, while in Australia the imported honeybee is rapidly exterminating the small, stingless native bee. One species of the charlock weed (*Sinapis arvensis*) (Figure 3.9) will commonly supplant another, and so on in other cases. We can dimly see why the competition should be most severe between allied forms, which fill nearly the same place in the economy of nature,[4] but probably in no one case could we precisely say why one particular species has been victorious over another in the great battle for life.

---

[4] If Darwin were writing today he would talk instead of species occupying the same niches.

Figure 3.9 Charlock weed (*Sinapis arvensis*).

(A)

(B)

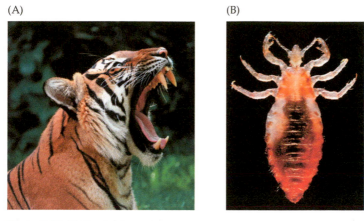

Figure 3.10 (A) The tiger's teeth are adapted to the types of prey that it catches for food. (B) The louse's legs are specialized for gripping the hairs on the tiger's body.

A corollary of the highest importance may be deduced from the previous remarks: namely, the structure of every organism is related, in the most essential yet often hidden manner, to that of all the other organisms with which it competes for food or residence, or from which it has to escape, or on which it preys. This is obvious in the structure of the teeth and talons of tigers and in the morphology of the legs and claws of the parasite that clings to the hair on the tiger's body (Figure 3.10). But in the beautifully feathered seed of the dandelion (members of the genus *Taraxacum*), and in the flattened and fringed legs of the water beetle (Figure 3.11), the morphological features might at first seem merely related to the art of living in air or water. Yet the true advantage of feathered seeds no doubt relates closely to the land being already thickly clothed by other plants; the feathering of the seeds allows them to be more widely distributed and to sometimes fall on unoccupied ground. In the water beetle , the

(B)

(A)

Figure 3.11 (A) Water beetle (*Dystiscus latissimus*). (B) Feathered seed of a dandelion.

structure of its legs, so well adapted for diving, allows it to compete successfully with other aquatic insects, to hunt for its own prey, and to escape being eaten by other animals.

The storage of large amounts of nutrients within the seeds of peas, beans, and many other plants seems at first sight to have no obvious relationship to interactions with other plants. But from seeing the strong growth of young plants produced from such seeds when sown in the midst of long grass, I suspect that the seed's nutrients are chiefly used to promote rapid growth of the young seedling as it struggles with the other plants growing vigorously all around it.

Competition and predation are powerful forces. Look at any plant in the middle of its range: Why does it not double or quadruple its numbers? We know that it can perfectly well withstand a little more heat or cold, or a little more dampness or dryness, for elsewhere it ranges into slightly hotter or colder areas, and slightly damper or drier areas. So what is holding it back? In this case we can clearly see that if we wished in our imagination to give the plant the power of increasing its numbers in any particular area, we should have to give it some advantage over its competitors, or over the animals that prey on it. On the edges of its geographical range, a change of constitution with respect to climate would clearly be an advantage to our plant; but we have reason to believe that only a few plants or animals range so far that they are destroyed by the rigor of the climate alone. Not until we reach the extreme confines of life, in the Arctic regions or on the borders of an utter desert, will competition with other individuals and other species cease. In most cases, even if the land is extremely cold or dry, yet there will still be competition between some few species, or between the individuals of the same species, for the warmest or the dampest spots.

Thus, we can also see that when a plant or animal is placed in a new area amongst new competitors, the conditions of its life will generally be changed in some essential manner even though the climate may be exactly the same as it was in its former home. If we wished to increase its average numbers in its new home, we would have to modify it differently here than if it was still in its native country; we would now need to do something to give it an advantage over a different set of competitors or help it to better deal with its enemies.

It is a good exercise to try in our imagination to give any particular organism some advantage over another. Probably in no single instance will we know what to do so as to ensure success. It will convince us of our ignorance of the mutual interactions of organisms in the wild, a conviction as necessary as it seems to be difficult to acquire. All that we can do is to keep steadily in mind that each living being is striving to increase at an exponential rate, and that each individual at some time in its life, during some season of the year, during each generation or at intervals, has to struggle for life, and that populations must periodically suffer great destruction. This seems terribly grim. But when we think about this great struggle, perhaps we can console ourselves with the full belief that the war of nature is not incessant, that no fear is felt, that death is generally prompt, and that the vigorous, the healthy, and the happy survive and multiply.

Key Issues to Talk and Write About

1. What does Darwin mean by "the struggle for existence"? Why does he claim that such struggles are inevitable?

2. According to Darwin, what factors control the number of individuals of any particular species that live in any particular area at any particular time?

3. Darwin talks about investigating the plants on one of his relative's estates (see page 53). According to Darwin, what accounts for the difference in species composition of the plants and insects on different parts of that estate? What is his evidence? How convincing do you find his conclusion?

4. Darwin argues that the presence of cats in a neighborhood might determine the abundance of flowers such as red clover (see page 56). Summarize his argument in a single sentence or two, as though you were explaining this to a friend or parent.

5. Why is Darwin convinced that struggles among individuals are typically the most intense between individuals of the same species, and between individuals of different species that are found in the same genus? Summarize his argument in your own words.

6. Find out two interesting things about one of the people that Darwin mentions in this chapter. Choose from the following:

   Sir Charles Lyell   Augustin Pyramus de Candolle
   Thomas Malthus   Herbert Spencer
   Carl Linnaeus   Félix de Azara

7. Take a look at one of the following two paragraphs:

   Page 47, the paragraph that begins, "When we look on the face of nature, seemingly bright with gladness…

   Page 53, the paragraph that begins, "We have many records showing how complex and often unexpected the checks…"

   For the chosen paragraph, what are the two or three main points that Darwin wishes to get across?

   Now summarize those points in a single sentence: your sentence should include all the major points, be accurate, make sense to someone who has not read the original paragraph, and be in your own words.

8. Here are three sentences from the original version of *The Origin of Species*.

   Rewrite each sentence to make it shorter and clearer.

   a) The causes which check the natural tendency of each species to increase are most obscure.

   b) Now the number of mice is largely dependent, as everyone knows, on the number of cats.

   c) I have also found that the visits of bees are necessary for the fertilization of some kinds of clover.

Online Resources *available at* sites.sinauer.com/readabledarwin

Videos

3.1 Time-lapse video of a dandelion forming seeds, and the seeds blowing away in the wind

3.2 Dandelion seeds being blown away by the wind

Links

3.1 Read about the five mass extinctions that have occurred on Earth in ancient times

science.nationalgeographic.com/science/prehistoric-world/mass-extinction/

3.2 Read an interview with Elizabeth Kolbert, who has published a book (2014) called *The Sixth Extinction* (Henry Holt and Co., publishers), about some of things that have been happening recently to natural populations of many organisms around the world

http://news.nationalgeographic.com/news/2014/02/140218-kolbert-book-extinction-climate-science-amazon-rain-forest-wilderness/

3.3 You can listen to another interview with Elizabeth Kolbert, courtesy of NPR

www.npr.org/2014/02/12/275885377/in-the-worlds-sixth-extinction-are-humans-the-asteroid

(Note: Web addresses may change. Go to sites.sinauer.com/readabledarwin for up-to-date links.)

4

Natural Selection,
or the Survival of the Fittest

In this chapter, Darwin emphasizes that natural selection works slowly and over extremely long periods of time, and that it works by favoring variations that provide advantages to the organisms that have them, in the struggle for nutrients and space and against predators and physical stresses, and even in the struggle for mates. Such "sexual selection" explains why individuals of one sex (usually the males) sometimes look so different from individuals of the other sex of the same species. Natural selection also works not just on adults but also on all other stages of development—eggs, seeds, embryos, larvae, and juveniles—and can act to either increase or decrease the complexity of various organs. In all cases, the best-adapted individuals will typically survive and propagate their offspring more successfully to the next generation, and those offspring will inherit the characteristics that made their parents so successful. Darwin also discusses the importance of outbreeding (matings among unrelated individuals) in promoting the vigor of offspring, and the negative impacts of self-fertilization and inbreeding (matings between close relatives). He also describes the evolutionary advantages of "divergence in character" over time—individuals that differ in certain ways from their fellows can exploit new niches and reduce competition with others of their own species, eventually driving intermediate forms to extinction. Finally, he provides a detailed example of natural selection in action, creating new species, new genera, and even new families and orders from a few ancestral species in a single genus, over many 1,000s of generations. A great many species go extinct along the way, because of their inability to compete with those newer creations that are better adapted to their conditions of life.

How will the struggle for existence, briefly discussed in the previous chapter, act on the variation among individuals? Can the principle of selection, which as we have seen in Chapter 1 is so very potent in human hands— apply equally well in nature? I think we shall see that it can indeed act in a very similar way, and most effectively.

In reading this chapter, keep in mind the seemingly endless number of slight variations and individual differences that occur in our domesticated animals and plants, and also in those in the wild, as well as the tendency for individual traits to be passed along to the owner's offspring. Remember, too, that the variability we see to so great an extent in our domesticated animals and plants is not directly produced by us: we can neither cause such varieties to occur nor prevent their occurrence; all we can do is preserve particular variations that do occur, and cause them to accumulate in offspring, generation after generation, as discussed in Chapter 1. And keep in mind how infinitely complex and closely linked are the mutual relations of all living things to each other in nature and to the physical demands and opportunities of their lives (see Chapter 3), and think how the infinitely varied diversities of structure within a population might be of use to individual organisms as the conditions around them change over time.

Considering how variations in the traits of individual plants and animals have been so useful to us—in farming, for example—how improbable can it be that other variations can also be useful in some way to each organism in the great and complex battle for life in nature? Some variations will surely be helpful to those individuals that have them. Remember that many more individuals are born in nature than can survive (see Chapter 3); can we doubt, then, that individuals having even slight advantages over others would have the best chance of surviving and successfully leaving offspring, and of passing those traits on to those offspring? On the other hand, it must also be true that any variation that is in the least harmful to its owner would be ruthlessly destroyed, through the owner's death or unsuccessful reproduction. **This gradual preservation of favorable individual differences and variations, and the destruction of those that are harmful, I have called natural selection, or "survival of the fittest."**

Some authors have misunderstood or objected to the term "natural selection," and some have even imagined that natural selection *induces* variation. **But no: natural selection only implies the preservation of variations that come about naturally, if those variations benefit the individual that possesses them.** No one objects to agriculturists talking about the potent effects of *man's* selection for particular traits; but remember that man can select for such particular traits only after such individual differences first occur in a population naturally.

Others have objected that the word "selection" implies conscious *choice* by the animals that become modified, as though the animals were doing the selecting. It has even been argued that as plants have no ability to make decisions on their own, natural selection cannot apply to them! But of course there is no deliberate choice here: selection is essentially imposed on the animals and on the plants. That is, individuals with certain characteristics are more likely to survive and leave offspring, many of which will inherit those characteristics, while individuals with certain other characteristics are more likely to die without reproducing.

Some have said that I speak of natural selection as an active power or deity; but who ever objects when an author says that the "attractions of gravity" rule the movements of the planets? Everyone knows what is meant and implied by such metaphors, and knows that they are almost always used for their brevity. Indeed, it's difficult to avoid personifying the word "nature." But when I say "nature" I mean only the aggregate action and product of many natural laws, and by "laws" I mean the sequence of events as ascertained by us. Such superficial objections to "natural selection" will eventually be forgotten, once people get used to the term.

We can best understand the probable course of natural selection over long periods of time by considering the case of a country undergoing some slight physical change—in climate, for example. Populations of most species will quickly change in size, and some species will probably go extinct. From what we have already learned about the intimate and complicated interactions that bind the organisms of each region together, it is clear that any change in the population size of some species will seriously affect all others in the area, independently of the direct effect of the change in climate itself. If the region were accessible at its borders, new forms would certainly immigrate into the area, and this would seriously disturb the interactions among some of the original inhabitants; I have already shown how powerful the influence of a single introduced tree or mammal can be on indigenous populations (see Chapter 3). On the other hand, in the case of an island or of some other parcel of land partially surrounded by other physical barriers that prevent new and better-adapted organisms from entering the area, climate changes would create new niches that would assuredly be filled up eventually if some of the original inhabitants, or their offspring, were in some manner modified; for if those areas had been open to immigrants from elsewhere those same new niches would have been seized upon by the intruders. In such cases, even slight modifications that better adapted the individuals of any particular species to the new climatic conditions would tend to be preserved and become more and more common over the generations. And natural selection would have free scope for the work of such improvement. Again, we see the importance of natural variation: without natural variation in traits, natural selection can do nothing. And please don't forget that the term "variation" includes *all* individual differences, not just those that are favorable to an organism. As we can produce great results with our domesticated animals and plants by adding up individual differences in any given direction over many generations, so can natural selection...but natural selection can do this even more easily, from having incomparably longer periods of time to bring such changes about.

Although climate change can certainly be a driving force for natural selection, I do not believe that it requires any great physical change—such as substantial shifts in climate or any unusually large degree of isolation (to prevent immigration)—for selection to work. For as all the inhabitants of any particular region are struggling together with nicely balanced forces, even extremely small modifications in the structure or habits of an individual over time could be enough to give it an advantage over others and increase the likelihood of its survival and reproduction; some of those slight modifications would then be passed on to at least some of its offspring. And still further modifications along the same lines would often still further increase the advantages for the offspring of that individual, as long as the species continued under the same conditions of life and profited by similar means of subsistence and defense.

There is no place on earth in which all native inhabitants are now so perfectly adapted to each other, and to the physical conditions under which they live, that none of them could become still better adapted or "improved" over the generations. Indeed, in all places that have been studied at least some native populations have been overwhelmed and conquered by immigrant species; the immigrants have taken firm possession of the land. As such invasive or introduced species have thus in every country successfully beaten some of the native species in the struggle for life, we may safely conclude that the native species could have been better adapted to their environment, so as to have better resisted the intruders. But they weren't; and they didn't.

As people *can* produce and certainly *have* produced many great results by our methodical and unconscious means of artificial selection among domesticated animals and plants (see Chapter 1), what changes might natural selection not bring about? People can only act on external characters that we can see and select for. But nature, if I may be allowed to personify the natural preservation or "survival of the fittest," cares nothing for mere appearances unless those appearances are useful to the individual in some way. And nature can act on every internal organ, on every physiological difference...on the whole machinery of life.

We select only for our own good. But nature always selects for the good of the individual that she acts on. And every selected character is fully acted on by her, as is implied by the fact of its selection. In our agricultural and animal domestication programs, we keep native species originating from many different climates together in the same place and we seldom exercise each selected character in any specific and fitting manner for it under domestication: we feed the same food to both long- and short-beaked pigeons; we do not exercise an unusually long-backed or long-legged quadruped[1] in any unusual manner; we expose sheep with both long and short wool coats to the same physical environments. Similarly, we don't allow the most vigorous males to struggle with each other for the females in mating; we select who will mate with whom. And we do not rigidly destroy all inferior animals each generation, but instead protect, as far as we can, each one during each varying season. We typically begin the selection process by selecting some modification prominent enough to catch the eye or be plainly useful to us. Under nature, though, the slightest differences of structure or

---

[1] Any four-legged animal, such as dogs, horses, and cows.

constitution may well turn the nicely balanced scale in the struggle for life, and so be preserved.

How fleeting are the wishes and efforts of man! How short his time! And consequently how poor will his results be, compared with those accumulated by nature during the whole of geological time! Can we wonder, then, that nature's productions should be far "truer" in character than our productions, and that they should be infinitely better adapted to the most complex conditions of life, and should plainly bear the stamp of far higher workmanship?

It may be said metaphorically that natural selection is scrutinizing—daily and even hourly—every variation throughout the entire world, rejecting those that are bad and preserving and adding up all that are good, silently and insensibly working whenever and wherever the opportunity allows, for the improvement of every living organism in relation to the physical and biological conditions of its life. We can see nothing of these terribly slow changes in progress until huge amounts of time have passed, and then so imperfect is our view into long-past geological ages that we see only that the forms of life are now quite different from what they had been.

In order for any species to become greatly modified over time, a new variety, once formed, must continue to show variation in individual characteristics among its members as before, so that helpful variations can be preserved in each generation, and so on over time. Seeing that individual differences in the same traits do indeed recur in each subsequent generation, this assumption of continued variation seems fully warranted. Whether or not it is in fact true we can judge only by determining the degree to which the hypothesis agrees with and explains what we see in nature.

Natural selection can act only for the good and benefit of individuals, but the characteristics it acts on need not be of obvious importance to us; characteristics and structures that seem to us to be of trifling importance can be acted on in the same way. When we see that leaf-eating insects are green, while insects feeding on tree bark are mottled gray, how can we not believe that these color patterns help protect these insects from their predators, by helping to camouflage them against their normal backgrounds (Figure 4.1)? Similarly, we see

(A)   (B)

Figure 4.1 (A) Leaf insect (*Phlogophora meticulosa*) camouflaged against a leaf. (B) Peppered moth (*Biston betularia*) camouflaged on an oak tree.

(A)

(B)

Figure 4.2 (A) Rock ptarmigan
(*Lagopus muta*) in winter plumage.
(B) Red grouse (*Lagopus lagopus*).

that the rock ptarmigan (Figure 4.2A) is white in winter, while the red grouse (Figure 4.2B) is the color of the heather in which it lives. Such coloration must be protective. Grouse, if not destroyed at some point in their lives would eventually form massive populations. But they are in fact known to be eaten, especially by birds of prey, such as the hawk. And hawks are guided to their prey by eyesight—so much so that in Europe, people are warned not to keep white pigeons, as these are most likely to be seen and eaten by hawks. Thus, natural selection should be extremely effective in eventually giving an appropriate color to every kind of grouse, and in keeping that color, when once acquired, true and constant, if in fact the color is beneficial to its possessor.

Nor should we think that the occasional destruction of an animal of any particular color would produce little effect on a population. People who raise sheep know how essential it is to destroy a lamb with the faintest trace of black in a flock of white sheep: one dark animal is all it takes to attract predators. We have also seen (Chapter 1) how the color of hogs in Virginia determines whether they shall live or die when feeding on the "paint-root" plant (*Lachnanthes tinctoria*). In plants, botanists typically think that the downy fuzz on the surface of fruit and the color of the flesh are characters of the most trifling importance; and yet we hear from that excellent horticulturist Andrew Jackson Downing (*The Fruits and Fruit Trees of America*, 1845) that smooth-skinned fruits in the United States suffer far more damage from curculio beetles (Figure 4.3) than those with external fuzz. Similarly, purple plums suffer far more from a certain disease than do yellow plums, while another disease attacks yellow-fleshed peaches far more than those with flesh of other colors. If such seemingly slight differences make such a great difference in cultivating the several varieties of fruit for us, then surely, in a state of nature, where trees would have to struggle with other trees and with a host of enemies, such differences would effectively determine which variety—whether a smooth or downy fruit, or a yellow or purple-fleshed fruit—would succeed.

In looking at many small points of difference between the structures of various species, some of which may seem quite unimportant and even trivial to us, we also need to bear in mind that when one part varies—and when those variations are accumulated through natural selection—then other modifications will also occur, sometimes of a very unexpected nature.

What else can we say about variation? Variations that appear at any particular period of life under domestication tend to reappear in the offspring at the same point in development—for example, in the shape, size, and flavor of the seeds of the many varieties of our plants used for cooking and agriculture; in the caterpillar and cocoon stages of the various varieties of silkworm; in the eggs of poultry, and in the color of the down on the skin of their chickens; and in the horns of our sheep and cattle

Figure 4.3 Plum curculio beetle (*Conotrachelus nenuphar*).

when nearly adult. Similarly, in the wild, natural selection will be able to act on and modify organisms at any stage of development and at any age, through the accumulation of variations profitable at that age, and by their future inheritance by the next generation at a corresponding age. If it benefits a plant to have its seed more and more widely disseminated by the wind, I can see no reason why this could not be affected through *natural* selection, in the same way that a cotton planter can increase and improve the down on the pods of his cotton trees by *artificial* selection. Natural selection may modify and adapt the larva of an insect to a variety of contingencies that are wholly different from those that concern the mature insect; and these modification may bring about, through correlation, changes in the structure of adults. Conversely, modifications in the adult may over the generations affect the structure of the larvae. But in all cases, natural selection will ensure that those modifications shall not be injurious: for if they were injurious, the species would become extinct.

Natural selection will modify structures in the young in relation to the parent, and of the parent in relation to the young. In social animals, like bees and ants, it will adapt the structure of each individual in ways that benefit the whole community, if the community benefits from those selected changes. Contrary to what I have read in various natural history writings recently, what natural selection cannot do is modify the structure of one species for the good of another species: it must first and foremost benefit the owner of the modification.

A structure used only once in an animal's life, but which plays an important role at that time, might still be modified to any extent by natural selection. For example, consider the great jaws possessed by certain insects, used exclusively for opening the cocoon after metamorphosis, or the hard tip at the beak of baby birds before they hatch, used for breaking out of the egg. I have seen reports that more of the best short-beaked tumbler pigeons perish in the egg than are able to hatch out of it successfully; pigeon fanciers must in fact help the birds escape from the egg. Now if for some reason it came to be advantageous for a full-grown pigeon to have a much shorter beak, the process of modification would be very slow, and there would simultaneously be the most rigorous selection for young, unhatched birds that had the most powerful and the hardest beaks for breaking out of the eggs; baby birds with weak beaks would be unable to hatch and would inevitably die. Or perhaps more delicate and more easily broken eggshells would be selected for instead; shell thickness is in fact known to vary, just as every other structure does.

Note that with all organisms there must also be a great deal of destruction by random chance. Such fortuitous destruction can't possibly have much of an influence on the course of natural selection, because by definition no selection is taking place when all organisms are equally likely to die. For instance, various animals eat a vast number of eggs or seeds every year at random; these could become modified by natural selection over time only if they varied in some manner that protected them from their enemies. Yet many of these eggs or seeds would perhaps, if they hadn't been eaten, have yielded individuals better adapted to their conditions of life than any of those that happened to survive. So again, a vast number of mature animals and plants, whether or not they be the best adapted to their environments, must be destroyed every year by purely accidental causes, which would not be in the least degree mitigated by certain changes of structure or constitution that would in other ways benefit the species. But let the destruction of the *adults* be ever so heavy, or again let the destruction of eggs or seeds be so great that only a hundredth or a thousandth of them survive to develop, yet of those that *do* survive, it will still be the best-adapted individuals that will tend to propagate their kind more successfully to the next generation. If the numbers are kept down dramatically by the causes just mentioned, as must often be the case, then nearly all individuals will die and natural selection will be powerless to bring about any changes in beneficial directions. But this in no way means that natural selection cannot be very efficient at other times and in other ways. We have no reason to suppose that all species always undergo modification and improvement at the same time in the same area.

Sexual Selection

Physical peculiarities often appear in males *or* females when organisms are raised under domestication, and those characteristics soon become hereditarily associated with that one sex. Thus it is entirely feasible that the two sexes of a species can become modified in different ways through natural selection. Indeed we know that this occurs, which leads me to say a few words about what I have called "sexual selection." This form of selection depends not on a struggle for existence, but rather on a struggle between the individuals of one sex—generally the males—for mating with the opposite sex. The loser of the struggle in this case does not die; it simply fails to leave offspring, or at most leaves very few offspring.

Sexual selection is therefore less vigorous than natural selection. Generally, the most vigorous males—those that are best adapted for their places in nature—will most likely succeed in mating and leave the most offspring.

Figure 4.4 Male stag beetle (*Lucanus cervus*).

But in many cases victory will depend not so much on general vigor, but rather on having special weapons that are found only on males. A hornless stag or a spurless rooster[2] would have a poor chance of leaving many offspring, because females select males based on the size and quality of those characteristics. Sexual selection, by always allowing the victor to mate and produce offspring, must surely lead to males with indomitable courage, spurs of great length, and wings of particularly great strength to strike opponents, in nearly the same manner as the brutal cockfighter produces especially strong and fierce roosters (gamecocks) by carefully selecting only the strongest and most aggressive males for mating in each generation.

I don't know how far down the scale of nature this law of battle extends, but certainly male alligators have been described as fighting, bellowing, and whirling around like Indians in a war dance, for the possession of females. Similarly, male salmon have been observed fighting all day long, and male stag beetles (Figure 4.4) sometimes bear wounds from the huge mandibles of other males. That inimitable observer of insect behavior, Monsieur Jean-Henri Fabre, has often seen the males of certain hymenopterous insects[3] fighting to possess a particular female while the female sits by watching, an apparently unconcerned beholder of the struggle who later shyly leaves with the winner, for mating.

The battles are perhaps most severe between the males of polygamous animals,[4] which accordingly often possess special weapons for fighting. The males of carnivorous animals are already well armed, but even there sexual selection may lead to special means of defense, as with the mane of male

---

[2] Male roosters, turkeys, and other members of the avian order Galliformes have a sharp, bony, rear-ward projection from the lower leg, called a spur. It is used for defense.

[3] Hymenoptera is an order of insects containing wasps, bees, and other species that have two pairs of membranous wings and an egg-laying organ in females (the ovipositor) that is specialized for stinging or piercing.

[4] Polygamous refers to a single male routinely mating with more than one female.

Figure 4.5 Male sockeye salmon (*Oncorhynchus nerka*) from the north Pacific Ocean.

lions, and the hooked jaws of the male salmon (Figure 4.5)—for the shield may be as important for victory as the sword or the spear.

The contest for mates among birds tends to be more peaceful. All those who have studied the subject believe that the males of many species use their singing to compete intensely with each other in attracting females. Some birds, including the rock thrush of Guiana and the birds-of-paradise of New Guinea (Figure 4.6) gather together in large groups while successive males show off for the females—with the most elaborate care—their gorgeous plumage. They also perform strange antics before the females who, after standing by as spectators, eventually choose the most attractive partner. People who have paid careful attention to birds in confinement well know that they often show clear individual preferences and dislikes. Sir Robert Heron, for example, has described how a particular peacock with feathers of an unusually mottled coloration was eminently attractive to all of the hens in his menagerie. I can't go into all of the details here, but if a human breeder can in a short time select for beauty and an elegant posture in his miniature chickens (called bantams), according to his standard of beauty, then I can see no good reason to doubt that female birds, by selecting the most melodious or beautiful males according to their standard of beauty, might produce—over hundreds or thousands of generations—a marked change in those characteristics. Some well-known laws regarding the plumage of male and female birds of particular species in comparison with the plumage of their young can partly be explained through the action of sexual selection on variations occurring at different ages, variations that are then transmitted to the males alone or to both sexes at corresponding ages. Unfortunately I don't have space here to discuss this topic in greater detail.

Figure 4.6 Male bird-of-paradise (*Paradisaea minor*). Most members of this family (the Paradisaeidae) are found in New Guinea and its neighboring islands.

I believe, then, that when the males and females of any animal species have the same general habits of life but differ in structure, size, color, or ornamentation, such differences have been mainly caused by sexual selection; that is, by individual males with certain specific characteristics having had, in many successive generations, some slight advantage over other males in gaining mates—either in their weapons, means of defense, or in their general charms—characteristics that they have then transmitted only to their male offspring. That doesn't mean that I would attribute *all* differences between the sexes to sexual selection. For example, we see in our domestic animals some peculiarities arising and becoming associated with the male sex alone that seem not to have been deliberately encouraged through selection by man. Similarly, it's hard to believe that the tuft of hair on the breast of the wild male turkey can be of any use, and I doubt that the female birds view it as attractive. Indeed, had the tufts appeared under domestication, they would have undoubtedly been called a monstrosity and been actively selected against.

Examples of Natural Selection, or the Survival of the Fittest, in Action

In order to clarify how I believe natural selection acts, I would like to give one or two hypothetical illustrations. Consider a wolf, which preys on a variety of other animals, capturing some by craft, some by strength, and some by fleetness. And let us also suppose that the swiftest prey—deer, for instance—had from some change in the surrounding countryside increased in numbers, or that other, slower prey had for some reason decreased in numbers, during that season of the year when the wolf had the greatest difficulty in finding food. Under such circumstances, wolves would have to focus all their attention on deer, and the swiftest and slimmest wolves would have the best chance of surviving and so be preserved (i.e., be selected for), as long as they remained strong enough to master their prey at this or some other time of the year, when they might be compelled to prey on animals other than deer. I can see no more reason to doubt that this would be the result than that human breeders should be able to improve the fleetness of their greyhounds by careful and methodical selection from generation to generation, or by that kind of unconscious selection that follows from each person trying to keep the best dogs without any thought of eventually modifying the breed, as discussed in Chapter 1. I may add that, according to Mr. James Pierce, author of "A memoir on the Catskill mountains" (1823), there are, in fact, two varieties of wolf inhabiting the Catskill Mountains in the United States—one with a light greyhound-like body shape, which pursues deer as prey, and the other more bulky, with shorter legs, which more often attacks the shepherd's sheep.

Note that in the previous illustration I spoke of the slimmest individual wolves having an advantage in that particular situation, and not of any single strongly marked variation having been preserved. Although I have long appreciated the great importance of individual differences, it wasn't until I read a very interesting article in the *North British Review* in 1867 that

I appreciated how rarely such single variations could be perpetuated to future generations. The author of this review gives an example using a pair of animals that produce 200 offspring during their lifetime, of which, from various causes of destruction, only two on average survive to reproduce in turn. This is a rather extreme estimate for most of the higher animals, but must be quite common for most of the lower forms. He then shows that if a single individual were born into the population that varied in some way that made it twice as likely to survive in comparison with individuals without that characteristic, the chances would still be very much against its survival: twice a very small probability is still a very small probability. Suppose that it did survive and reproduce, and that half of its offspring inherited the favorable variation. Even so, as the article goes on to show, the young would still have only a very slightly better chance of surviving and breeding, and this chance would continue to decrease in the succeeding generations.

I can't dispute the logic of those arguments. If, for instance, a bird of some kind could obtain its food more easily by having its beak curved, and if one were born with its beak strongly curved and flourished as a consequence, even so there would be a very poor chance that this one individual would survive long enough to mate and thus to pass this characteristic on to its offspring. On the other hand, there can be little doubt, judging from what we see taking place when humans selectively breed our domesticated animals and plants, that such beneficial characteristics would indeed follow from the preservation during many generations of *a large number* of individuals with heritable, strongly curved beaks, and from the more common destruction of a still larger number of birds with the straightest—and therefore most disadvantageous—beaks.

Some have also suggested that mating between individuals with different characteristics will eliminate variations of all kinds in the next generation. I will have more to say about this later, but here let me just note that most animals and plants stay within a small area and do not needlessly wander about. We see this even with migratory birds, which almost always return to the same spot every year. Consequently, each newly formed variety would generally be at first local, as seems to be the common rule with varieties living in nature. Thus, similarly modified individuals would stay together in a small group and would often breed together. If the new variety was successful in the battle for life and for mates, it would then slowly spread from the central areas, competing with and conquering the unchanged individuals on the edges of an ever-increasing circle.

Let me give another, more complex illustration of how natural selection works, this time concerning plants. Certain plants excrete a sweet juice from specialized glands, a juice that apparently eliminates something injurious from the sap. Some legumes, for example, secrete such juices from specialized glands at the base of the stipules, while the common laurel has similar glands on the back of its leaves. Although the glands produce only small amounts of this juice, it is greedily sought by insects, whose visits have no apparent benefit for the plants.

Let's suppose that this same juice eventually came to be secreted from inside the flowers of a certain number of plants of some particular species. Insects would now get dusted with pollen while seeking the nectar, and would then often end up transporting that pollen from one flower to another, so that the flowers of two individuals of the same species would get mated, or "crossed." We know that such crossing gives rise to especially vigorous seedlings, which consequently would have the best chance of flourishing and surviving. The plants that produced flowers with the largest juice glands (often called "nectaries") would secrete the most nectar and thus be most often visited by insects. They would in consequence most often be crossed with the flowers on other plants, and so in the long run would gain the upper hand against other local members of the species. Eventually they would form a local variety. Any individual flower that had its stamens and pistils placed so as to favor in any degree the transport of pollen by the particular insects that visited them would likewise be favored.

I could instead have used an example of insects visiting flowers to feed on pollen rather than nectar. As pollen is produced solely for the fertilization of eggs, its destruction by visiting insects would seem to be a clear and simple loss to the plant. And yet, if even just a little pollen were carried from flower to flower by the pollen-devouring insects, at first only occasionally and later more regularly, thus resulting in a cross between plants, then it might still be a great gain to the plant to be thus robbed of its pollen, even if 90 percent of the pollen were eaten by the insects beforehand. Those individuals that produced more and more pollen, and had larger anthers for its production, would be selected for and come to be increasingly common in the population over many generations. If this process continued for long periods of time, then once our plant had become highly attractive to insects, those insects would routinely carry pollen from flower to flower, and without at all intending to do so. Indeed that is exactly what they do. I will give just one example, one which also illustrates a step in the separation of the sexes in plants.

Some holly trees bear only male flowers, each of which has four stamens (Figure 4.7) producing only a little pollen, and a rudimentary female portion, the pistil. Other holly trees bear only females flowers, with a full-sized pistil, but they also have rudimentary male parts: four stamens with shriveled anthers, in which not a grain of pollen can be detected. Having found a female holly tree exactly 60 yards from a

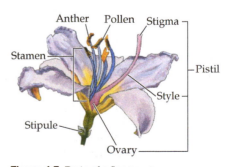

Figure 4.7 Parts of a flower.

male tree, I looked at the stigmas of 20 flowers, taken from a number of different branches, under the microscope; on every single one of these flowers I saw at least a few pollen grains, and on some a profusion of grains. As the wind had been blowing for several days in the opposite direction, from the female tree to the male tree, the pollen could not have been carried to the female by the wind. Even though the weather had been cold and boisterous, and therefore not especially favorable to bees, every female flower that I examined had been fully fertilized by the bees, which had flown from tree to tree in search of nectar.

But returning now to our imaginary case, as soon as the plant had been rendered so highly attractive to insects that pollen was regularly carried from flower to flower, another process might begin. No naturalist doubts the advantage of what has been called the "physiological division of labor." Thus we may believe that it would be advantageous for a plant to become more specialized and produce male parts alone—stamens—in one flower or on one whole plant, and female parts alone—pistils—in another flower or on another plant. When domesticated plants are placed under different environmental conditions, sometimes the male organs and sometimes the female organs do indeed become more or less impotent. Now if we suppose this to occur in ever so slight a degree in nature, then as pollen is already carried regularly from flower to flower, and as a more complete separation of the sexes of our plant would be very advantageous on the principle of specialization and division of labor, then those individuals showing this tendency more and more would be continually favored or selected for in each generation, until at last a complete separation of the sexes might be achieved. I don't have the space here to show the various steps through which plants of various types are now in the process of separating the sexes, but I may add that according to Dr. Asa Gray, the great American botanist at Harvard University, some species of holly in North America are in fact presently in exactly such an intermediate condition.

Let us now consider the nectar-feeding insects. Let's suppose that the plant for which we have slowly been increasing the nectar production through continued selection is a common one, and that certain insects depend in large part on that plant's nectar for food. I could give many details showing how anxious bees are to save time when obtaining nectar. For example, they have a habit of cutting holes and sucking the nectar at the bases of certain flowers, which, with a very little more trouble, they could have entered by the "mouth" of the flower. Bearing such facts in mind, it is easy to believe that under certain circumstances individual differences in the curvature or length of the bee's proboscis, too slight to be even noticed by us, might benefit a bee or other insect, so that certain individuals would now be able to obtain their food more quickly than others. Thus the communities to which they belonged would flourish and throw off many swarms inheriting the same peculiarities. Similarly, the tubes of the corolla of the common red and crimson clovers (*Trifolium pratense* and *T. incarnatum*, respectively) (Figures 4.8) do not obviously differ in length; and yet the honeybee (members of the genus *Apis*) can easily suck the nectar out of the crimson clover, but not out of the common red clover, which is instead visited only by bumblebees (genus *Bombus*). Thus entire fields of red clover offer in vain an abundant supply of precious nectar to the honeybee. And it's not that the honeybee dislikes the nectar of these flowers, for I have repeatedly seen, in the autumn, many honeybees sucking nectar from them through holes previously bitten in the base of the tube by bumblebees. The difference in the length of the corolla in the two kinds of clover—which determines whether or not honeybees can visit—must be very trifling, for I have been assured (although I have not myself made such an observation) that when red clover has been mown, the flowers of the second crop are somewhat smaller; honeybees can then successfully visit these flowers for their nectar. I have also been told that the Ligurian bee, which is usually considered to be a mere variety of the common honeybee and which freely breeds with it, is able to reach and suck out the nectar of the red clover. Thus in a region where this kind of clover was abundant, it might be a great advantage for honeybees to have a slightly longer or differently built proboscis. On the other hand, as the fertility of this red clover absolutely depends on bees visiting their flowers, if bumblebees were to become rare in any particular region it might be a great advantage to the plant to have a shorter or more

(A)

(B)

Figure 4.8 (A) A bumblebee feeds on a red clover (*Trifolium pratense*). (B) Honeybee on crimson clover (*Trifolium incarnatum*).

deeply divided corolla, so that the honeybees should now also be able to suck nectar from the flowers. Thus I can understand how a flower and a bee might slowly become, either simultaneously or one after the other, modified and adapted to each other in the most perfect manner, simply by the continued preservation of all those individual plants and bees in each generation that presented slight deviations of structure that were mutually favorable to each other.

I am well aware that my doctrine of "natural selection," as illustrated in the above imaginary examples, is open to the same objections that were initially leveled at the geologist Sir Charles Lyell's noble views on "the modern changes of the earth, as illustrative of geology." But we now seldom hear people talking about erosion and uplifting—agencies that we still see at work on this planet—spoken of as being trifling or insignificant when used to explain the excavation of the deepest valleys on our planet or the formation of long lines of inland cliffs. **Natural selection acts only through the preservation and accumulation of small inherited modifications, each in some way beneficial to the preserved individual. Just as modern geology has almost banished such ideas as the excavation of a great valley by a single, ancient, huge wave of water, so will the principle of natural selection eventually banish anyone's belief in the continued creation of entirely new organisms, or of any great and sudden modification in their structure.**

On the Outbreeding[5] of Individuals

Here I must briefly change the subject a bit and say something about the importance of matings between individuals that are not closely related to each other.

In the case of animal and plant species with separate sexes, it is easy to see why two individuals must usually come together to fertilize the eggs. The only exception is the curious and little-understood case of parthenogenesis, in which eggs can develop normally without being fertilized by sperm. But many organisms are hermaphrodites, with each individual having both male and female reproductive organs. For such hermaphroditic species, it is not at all obvious why two individuals must be involved in getting the eggs fertilized: Why shouldn't each individual simply fertilize its own eggs? And yet it seems that even among hermaphrodites, two individuals do at least occasionally come together for reproduction, as first suggested more than 50 years ago by the botanists Christian Konrad Sprengel, Thomas Andrew Knight, and Joseph Gottlieb Kölreuter. We will presently see in some detail why this is so important, but for now I will be brief. Among insects, vertebrates, and some other large animal groups, two individuals generally come together to produce offspring, even in hermaphroditic species. Yet many other hermaphroditic animals do not routinely pair, and the vast majority of plants are in fact hermaphrodites. What evidence do we have that two hermaphroditic individuals ever get together to reproduce?

Based on my extensive reading and many of my own studies, it is clear that crosses between different varieties of plants and animals greatly increase the vigor and reproductive capacity (i.e., the fertility) of their offspring, something that is also well-known by most professional breeders. The same result holds for matings between individuals of the same variety but belonging to different strains. It is also well understood that matings

---

[5] Outbreeding, called "intercrossing" by Darwin and sometimes referred to as "outcrossing," refers to matings between individuals that are not closely related to each other, thus increasing genetic diversity within the lineage. Inbreeding, in contrast, refers to matings between closely related individuals.

among closely related individuals ("inbreeding") reduces both vigor and fertility in the offspring. These facts alone make me think that it is a general law of nature that no living organism routinely fertilizes its own eggs generation after generation; crosses with another individual seem indispensable, even if they occur only infrequently.

Assuming that this is indeed a law of nature, we can, I think, understand several large classes of facts that would otherwise be inexplicable. For example, every plant hybridizer knows that exposing flowers to moisture reduces the chances of successful pollination (probably through the loss of pollen).

Why then do such a multitude of flowers have their anthers and stigmas fully exposed to the weather? And why do the plant's own anthers and pistil stand so near each other as almost to insure self-fertilization? Such placement of these flower parts can really make sense only if it is essential for flowers to at least occasionally cross with other flowers: having the anthers and stamens fully exposed to the air increases the likelihood that one flower's pollen will travel elsewhere, and that the eggs of one flower will be fertilized by pollen from another.

On the other hand, many flowers have their reproductive organs well enclosed, as in the great "pea family" (the Fabaceae, formerly known as the Leguminosae), and so cannot be pollinated by wind. Such flowers, however, almost inevitably present beautiful and curious adaptations that promote visits by insects. Indeed, so necessary, for example, are the visits of bees to peas and many other members of this great family that preventing such visits greatly diminishes the plants' reproductive output. It is essentially impossible for insects to fly from flower to flower without carrying pollen from one flower to the other, to the great indirect benefit of the plant. I can ensure fertilization of such flowers just by touching a painter's camel's-hair pencil first to the anthers of one flower and then to the stigma of another; bees and other insects must act in the same way, as they fly from flower to flower. This does not in any way suggest that bees produce a multitude of hybrids between distinct plant species, for if a plant's own pollen is placed on a stigma with pollen from another plant species, the plant's own pollen invariably destroys the influence of the foreign plant's pollen, as shown by the German botanist Karl Friedrich von Gärtner.

Intriguingly, flowers of the common barberry plant show a curious behavior that would seem to ensure self-fertilization: the stamens of the barberry flower either suddenly spring toward the pistil of the same flower, or they move toward it one after another. However, insects are often required to bring this springing behavior about, as shown by Mr. Kölreuter, and insects, as we have already noted, are excellent vehicles for transferring pollen from flower to flower. Indeed, although the members of this plant genus seem well-designed for self-fertilization through this clever contrivance, it is well known that if closely related forms are planted near each other it is almost impossible to raise pure seedlings, so often do they naturally cross.

In many other cases, not only is self-fertilization not promoted, but there are special contrivances that effectively prevent the stigma from receiving pollen from the same flower. This has been documented in the work of Mr. Sprengel and other botanists, and I have seen it myself. For example, in the cardinal flower (*Lobelia cardinalis*; see Figure 3.7A), there is a beautifully elaborate contrivance through which all of the incredibly numerous pollen grains are swept out of the fused anthers of each flower before the stigma of that particular flower is ready to receive them, thus nicely preventing self-pollination; as this flower is never visited by insects, at least in my garden, it never sets seed naturally. And yet, I can raise plenty of seedlings by simply transferring pollen from one flower to the stigma of another. A different species of *Lobelia* is visited by bees in my garden and seeds regularly without any help from me.

In many other cases, though, there is no special mechanical contrivance to prevent the stigma of one flower from receiving pollen from the same flower. And yet, as a number of botanists, including Mr. Sprengel, and more recently Friedrich Hildebrand, have shown, and as I can confirm from my own studies, either the anthers burst and release their pollen before the stigma of that same flower is ready to be fertilized, or the stigma is ready to receive pollen before the pollen grains of that same flower are ready to be released; thus these plants in fact have functionally separated sexes and must be habitually crossed, even though they are, strictly speaking, hermaphroditic. It is the same with the reciprocally dimorphic and trimorphic plants previously alluded to, in which each plant bears two or three distinctly different sorts of flowers with the styles and stamens at different heights: one flower will, for example, have a short style and long stamens, while another flower will have a long style and short stamens. Such flowers must be crossed if they are to produce seeds. How strange are these facts! And how strange that the pollen and stigmatic surfaces of the same flower, though found so close together as if to insure self-fertilization, should be in so many cases mutually useless to each other! But how simple it is to explain these facts if we understand the advantages, or even the indispensability, of at least occasional crosses with other individuals.

If several varieties of some plants, including the cabbage, radish, and onion, are allowed to seed near each other, most of the seedlings thus raised turn out to be combinations of the varieties—mongrels—as I have found through experiments in my own garden. For example, I raised 233 seedling cabbages from a number of plants of different varieties growing near each other; only 78 of those seedlings were true to their variety, and some of those were not even perfectly true. Yet the ovary-containing pistil of each cabbage flower is surrounded not only by its own six pollen-bearing stamens, but also by those of the many other flowers on the same plant. Moreover, the pollen of each flower readily gets on the stigma of the same flower without any help from insects; indeed, I have found that plants carefully protected from visits by insects produce the normal number of pods. How, then, is it possible that such a vast number of the seedlings are mongrels? It must arise from the pollen of a distinct variety having a greater potency than the flower's own pollen.[6] This is part of the general law that I discussed earlier, that substantial benefits must derive from outbreeding among distinct individuals of the same species. When distinct *species* are crossed the case is reversed: a plant's own pollen is almost always more potent than the pollen coming from flowers of a different species. I will return to this subject in a later chapter (Chapter 9).[7]

In the case of a large tree covered with innumerable flowers, it may be objected that pollen could seldom be carried from tree to tree, and at most only from flower to flower on the same tree, thus promoting self-fertilization since flowers on the same tree can be considered as distinct individuals only in a limited sense. I believe that this objection is valid, but that nature has largely provided against such self-pollination by giving to trees a strong tendency to bear flowers with separated sexes. When the sexes are thus separated, although male and female flowers may be produced on the same tree, pollen must be regularly carried from flower to flower for fertilization to occur; this will increase the likelihood of pollen being at least occasionally carried from tree to tree. I recently determined that in England, trees belonging to all orders have their sexes more often separated than other plants do.

Similarly, at my request Dr. Joseph Hooker, of the Royal Botanical Gardens, tabulated the trees of New Zealand, and Dr. Gray those of the United States, and the results were similar. On the other hand, Dr. Hooker informs me that the rule does not hold good in Australia. But if most of the Australian trees are dichogamous (i.e., having pistils and stamens that mature at different times), the same result would follow as if they bore flowers with separated sexes. I have made these few remarks on trees simply to call attention to the subject.

Let us turn briefly now to animals. Various terrestrial species, such as land snails and earthworms, are hermaphroditic. However, all of these animals pair for reproduction; I have yet to find a single terrestrial animal that can fertilize its own eggs. This remarkable fact, which offers so strong a contrast with terrestrial plants, makes perfectly good sense if an occasional cross with other individuals is indeed indispensable; there is nothing analogous to the action of insects or wind on plant reproduction by which an occasional cross could be achieved for terrestrial animals; it always requires the meeting of two individuals.

Self-fertilizing hermaphrodites are common among aquatic animals; but since they release their sperm into the water, water currents offer an obvious means for least an occasional cross between individuals. As in the case of flowers, I have so far failed—even after consulting Professor Thomas Henry Huxley, one of the highest authorities in biology—to discover a single

---

[6] Today we call this phenomenon "self-incompatibility."

[7] Darwin's cross-references to chapters beyond Chapter 8 have been retained in this volume so that readers can refer to the original *The Origin of Species*, Sixth Edition, if they choose.

hermaphroditic animal with its organs of reproduction so perfectly enclosed that it would be impossible for another individual to access them on occasion. For a long time, barnacles seemed to be a possible exception to this general belief, but I have now been able to prove, through a bit of good luck, that even here, two barnacles do sometimes cross with each other, even though both are normally self-fertilizing hermaphrodites.

It must have struck most naturalists as a strange anomaly that among both animals and plants, some species are hermaphrodites while others within the same family (and some even within the same genus!) have separate sexes (i.e., they are dioecious), even though they are extremely similar in all other aspects of their biology. But if, in fact, all hermaphrodites do at least occasionally intercross, the difference between hermaphroditic and dioecious species is, as far as function is concerned, very small.

From these several considerations and from the many other facts that I have collected but don't have sufficient space to discuss here, it appears that with both animals and plants at least an occasional intercross between distinct individuals is a very general, if not universal, law of nature.

Circumstances Favoring the Production of New Forms Through Natural Selection

This is an extremely intricate subject. A great amount of variability among individuals will obviously favor the process of natural selection by offering traits to be selected for or against. Having a large number of individuals in a population is, I believe, also highly important for success, by making it more likely that useful variations will appear within any given period, and compensating for lesser amounts of variability within each individual. Though nature grants long periods of time for natural selection to work, she does not grant an indefinite period: as all living beings are competing with each other for space, shelter, and food, if any one species does not become modified and improved over time as much as its competitors are so modified and improved, it will go extinct. And of course unless favorable variations are inherited by at least some of each individual's offspring, natural selection can accomplish nothing.

In the case of a human breeder methodically selecting for particular traits, his work will completely fail if the individuals are allowed to freely intercross with each other. But when many men, without intending to alter the breed at all, have a very similar standard of perfection in mind that they all work toward, and all try to procure and breed from only the best of their animals, improvement surely but slowly follows from this unconscious process of selection even when no special effort is made to prevent interbreeding with inferiors.

It will be the same in nature; for within any confined area if there is some niche not already perfectly occupied by some animal or plant species, all individuals varying in the proper direction, regardless of degree, will more likely be preserved and reproduce. But if the area is large, its various districts will almost certainly present different environmental conditions; if so, any newly formed varieties will interbreed with others on the edges of each such district. But we shall see in Chapter 6 that intermediate varieties, living in intermediate districts, will in the long run generally be supplanted by one of the neighboring varieties.

Outbreeding will chiefly affect those animals that unite for each birth and wander about a good deal, and which do not reproduce very quickly. Thus with birds and other animals of this nature, particular varieties will generally be confined to rather large areas that are particularly far apart and at least somewhat isolated, located, for example, in separated countries. Indeed, I find this to be the case. With hermaphrodites and organisms that unite for each birth but move about less and leave many offspring, a new and improved variety might be quickly formed in any small area, and might first maintain itself there within a small group and later spread further, so that members of the new variety would mate mostly with each other. Thus we are likely to see more varieties occurring over much smaller distances. Following this principle, nurserymen always prefer to save the seed produced by a large number of plants of any particular variety, thus reducing the chance of outcrossing with other varieties.

Even with animals that come together for each birth and that do not increase their numbers quickly, we must not assume that free outbreeding among individuals would always take place and eliminate the effects of natural selection. Indeed, I now have a considerable body of facts showing that within the same area, two varieties of the same species may long remain distinct from each other, just by living in slightly different places within that area or from breeding at slightly different seasons, or from the individuals of each variety preferring to mate with others within that variety.

Outbreeding is important in nature, in keeping individuals of the same species, or the same variety, true and reasonably uniform in character over time. Obviously it will act far more efficiently with those animals that unite for each birth, but, as I stated earlier, there is good reason to believe that all animals and all plants exhibit outbreeding at least on occasion. Even if this happens only infrequently, the young thus produced will gain so much in vigor and fecundity compared with the offspring from long-continued self-fertilization, that they would have a better chance of surviving and then propagating their kind. Thus the influence of crosses between individuals will always be great, even if they happen rarely.

Isolation also plays an important role in allowing species to be modified through natural selection. In any small isolated area, the conditions of life will generally be almost uniform; thus natural selection will tend to modify all of the varying individuals of any one species in the same direction. This will deter crossbreeding with inhabitants of surrounding areas. The German explorer and natural historian Moritz Friedrich Wagner has recently shown that the role of isolation in preventing crossbreeding

84   Chapter 4

between newly formed varieties is probably even greater than I had imagined, although for reasons already noted, I disagree with his suggestion that physical isolation is essential for the formation of new species.

Isolation is also important in preventing potentially better adapted organisms from immigrating into an area after environmental or physical conditions have changed—following a change in climate, for example, or a change in elevation of the land. Thus any newly created niches will remain open, to be filled eventually by the gradual selection of suitable variants from descendants of the original populations. Finally, isolation from other areas will allow plenty of time for a new variety to be improved gradually, at a very slow rate. On the other hand, to be effective at promoting evolution by natural selection, the isolated area cannot be too small, for then the total number of inhabitants living in that area will be small as well; this will greatly slow the production of new species through the process of natural selection, by decreasing the chances of favorable variations arising in the population in the first place: if there is no variation, there is nothing to be selected for or against.

Note that the mere passage of time alone does not guarantee that species will be modified. Organisms do not change their traits automatically over time, through any innate law. The passage of time simply increases the likelihood that beneficial variations will eventually arise within a population, and be selected for, accumulate, and eventually become fixed.

Now let us see how all of these remarks about isolation apply in nature, by considering what can happen on any particular small, isolated area such as an oceanic island. Although only a small number of species will be found living on that island, as we shall see in a later chapter on Geographical Distribution (Chapter 13), a very large proportion of those species will be found only there and nowhere else. Thus oceanic islands might at first seem especially likely to produce new species. We shouldn't jump to that conclusion though: we really can't know whether small isolated areas like islands, or large open areas like continents have been most favorable for producing new varieties and species unless we know the amount of time taken to do so in each case…and this is something that we simply do not know.

Although isolation is very important in producing new species, I think that the largeness of the area involved is even more important, especially for producing species that can endure for long periods of time and spread widely. Not only will favorable variations be more likely to appear in a large and open area, simply because of the greater number of individuals living in such areas, but the diversity of habitats and niches will also be more complex in such an area because of the large number of species that already live there. If some of those many species become modified and improved over time, that will favor modification and improvement in the species that they interact with as well; those that fail to improve under such pressure will eventually go extinct.

Each new form, once it has become substantially improved, will be able to spread over the large open and continuous area, and thus will come into competition with many other forms. Moreover, areas that are now large and continuous will often have been substantially fragmented in the past, owing to previous rises and declines in sea level, and in the elevation or subsidence of land, and thus have benefitted from the helpful effects of isolation long ago.

Finally, my reasoning suggests that although small isolated areas have in some respects favored the production of new species, yet such modification will generally have been quicker on larger areas like continents. Even more importantly, the new forms produced on larger areas of land will already have been victorious over many competitors, and will thus spread the most widely and give rise to the greatest number of new varieties and species; accordingly, they will play a more important role in the changing history of the organic world.

This superior competitive ability of forms that have developed on large continents is clearly seen in a number of situations. The species of the smaller continent of Australia, for example, are now being outcompeted by those introduced from the larger European and Asiatic areas, as discussed more fully later in my chapter on Geographical Distribution (Chapter 12). The superior competitive ability of species produced on large continents also explains why such species have so often become dominant after being introduced onto islands. On a small island, competition will have been less severe, and consequently there will have been less subsequent modification and extinction. We can understand, then, why the flora presently found on the island of Madeira resembles, according to the Swiss paleobotanist Oswald von Heer, the now extinct European flora from the Tertiary period (about 65 million to 2 million years ago): weaker forms that have persisted on the islands have been outcompeted to extinction on the larger continents.

Similarly, competition among freshwater organisms should also have been relatively less severe than elsewhere, because all freshwater basins taken together make a far smaller area than that of the ocean or the land. Consequently, without the impetus of severe competition, new forms should have been produced more slowly in freshwater environments, and gone extinct more slowly as well. Not surprisingly then, in freshwater habitats we find seven genera of primitive ganoid fishes, with their odd, thick bony scales, the last remnants of a once far-more-common order. And it is in freshwater, too, that we find some of the strangest forms known in the world: the platypus (Figure 4.9A), which belongs to the mostly extinct family Ornithorhynchidae, and the lungfish (Figure 4.9B), a member of the order Lepidosireniformes. In marine habitats, most such forms have been driven into extinction. These anomalous freshwater forms can be considered as "living fossils"; they have survived to the present day by inhabiting a confined area, and from having been exposed to a smaller variety of—and therefore less severe—competitive interactions.

Let me now summarize the circumstances that favor, and those that impede, the production of new species through the process of natural selection.

(A)   (B)

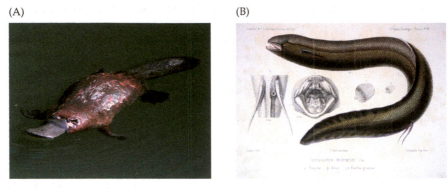

Figure 4.9 (A) Australian duck-billed platypus swimming. (B) South American lungfish (*Lepido-siren paradoxa*). In this and in most other lungfish species, the gills are incapable of substantial gas exchange; instead, each fish has two lungs and breathes air. The Australian lungfish has only a single lung, but is still able to use its gills in respiration.

For terrestrial species it seems that a large continental area, and one that has undergone many oscillations of level, will have been most likely to have produced many new life forms, fitted to endure for a long time and to spread widely. While the area was part of a continuous large continent, there will have been many kinds of individuals and many individuals of each kind, creating severe competition among the various forms. When these large areas were converted by the sinking of land or the rising of sea level into separate large islands, each of those islands will have supported many individuals of the same species. Matings between individuals on the edges of each species range will have been impossible, and after physical changes of any kind, immigration from the outside would not have been possible either. Thus any new niches opening up on each island would have to have been filled through the gradual modification of the original inhabitants, and there will have been sufficient time for the varieties in each to have become well modified and perfected. When the islands were later reconverted into a large continuous continental area, by elevation of the land or a lowering of sea level, competition will again have become intense: the best adapted and most improved varieties will have been able to spread, while the less improved forms will have been outcompeted to extinction. The relative proportions of the various forms on the reconstituted continent will again have been changed. And once again there will have been a fair field for natural selection to act, in improving the inhabitants still further and thus in producing new species.

I fully admit that natural selection generally acts extremely slowly, and can act only when some niches in an area can be better occupied by the modification of some of the existing inhabitants of that area. The creation of such areas will often depend on physical changes, which generally take place very slowly, and on the immigration of better adapted forms into the area being in some way prevented. As a small number of individuals become modified, the interactions with other inhabitants will often be disturbed. This will in turn create new niches, ready to be filled up by better adapted forms. But all of this will take place very slowly. Although every individual differs in some slight degree from other members of its species, it would often be a long time before differences of the right sort in particular traits appeared in a population. And such progress would often be greatly slowed by free crossbreeding.

Many will exclaim that these several limitations are sufficient to neutralize the power of natural selection. I, however, do not believe so. But I do believe that natural selection will generally act very slowly, and only on a few of the inhabitants of any given region. I further believe that these slow, intermittent results agree well with what geology tells us about the rate and manner at which the inhabitants of the world have in fact changed over long periods of time.

Slow though the process of natural selection may be, if feeble humans can accomplish as much as we have by artificial selection, I can see no limit to the amount of change, or to the beauty and complexity of the co-adaptations between all organic beings, one with another and with their physical conditions of life, which may have been brought about in the long course of time through nature's power of selection—that is, through survival of the fittest.

Extinction Caused by Natural Selection

I will discuss this subject more fully in a later chapter on geology (Chapter 11), but as extinction is intimately connected with natural selection I must say something about it here as well.

Natural selection acts solely by preserving variations that are in some way advantageous to the organisms that possess them, and which consequently endure. Owing to the exponential rate of increase of all living organisms, each habitat is already fully stocked with inhabitants. It follows, then, that as the favored forms increase in number, so, generally, will the less favored forms decrease, and eventually become rare. **Rarity, as geology tells us, is the precursor to extinction.** We can see that when any form is represented by only a few individuals it will run a good chance of complete extinction, for example during a time of great climate fluctuation, or from a temporary increase in the number of its enemies. But we may go further than this: as new forms are produced, many old forms *must* become extinct, unless we assume that specific forms can increase in numbers indefinitely. But geology plainly tells us that the total number of species has not increased indefinitely over time. I shall presently attempt to show why the number of species throughout the world has not become immeasurably great.

We have seen that species with the greatest number of individuals also have the best chance of producing favorable variations within any given period; if only a very small percentage of variations are favorable, then such variations are not likely to appear in populations with only a small number of individuals. And I showed in Chapter 2 that it is the most common and widespread or dominant species that presently show the greatest number of recorded varieties. Thus species with relatively few individuals (the so-called "rare species") will be less quickly modified or improved within any given period, and in consequence will probably be beaten in the race for life by the modified and improved descendants of the more common species.

From these several considerations, I think it inevitably follows that as new species are formed over time through natural selection, others will become rarer and rarer, and finally go extinct. The closest competitors with those undergoing modification and improvement will naturally suffer the most. And as seen in the chapter on The Struggle for Existence (see Chapter 3), it is the most closely allied forms—varieties of the same species, and species of the same genus or of related genera—that, from having nearly the same structure, physiologies, and habits, generally come into the severest competition with each other. In consequence, each new variety and each new species, during the progress of its formation, will generally press hardest on its closest relatives and tend to exterminate them.

We see the same process of extermination among our own domesticated animals and plants, through the selection of improved forms by humans. I could give many curious instances showing how quickly new breeds of cattle, sheep, and other animals—and new varieties of flowers as well—completely take the place of older and inferior kinds that simply have appealed less to their human breeders. In Yorkshire, England, we know with certainty that the ancient black cattle were displaced by longhorn cattle, and that these were, as noted so nicely by William Youatt, "swept away by the short-horns as if by some murderous pestilence." Extinction is an inevitable part of the process of selection.

Divergence of Character

The principle of "divergence of character" is extremely important to my argument. I believe it explains several important facts. In the first place, although varieties—even strongly marked ones—may have somewhat the character of species, they differ far less from each other than do good and distinct species. But if varieties are incipient species, as I have suggested— new species in the making—then how does the smaller difference between varieties eventually become enlarged into the greater difference between species? That this does routinely happen we must infer from this fact: most of the innumerable species throughout nature are very clearly different from each other, whereas varieties—the supposed prototypes and parents of future well-marked species—present only slight and ill-defined differences. Mere chance might cause one variety to differ in some character from its parents, and by chance cause the offspring of this variety again to differ from its parents in the very same character and to a greater degree; but this alone would never account for so habitual and large a degree of difference as that seen between species of the same genus. So how can we explain this transformation of a variety into a new species?

As usual, I have turned to our domestic animals and plants for some help in answering this question. The situation is fully analogous. Clearly, the production of races so different from each other as, for example, short-horn and Hereford cattle, or racehorses and cart horses (Figure 4.10), or the various breeds of pigeons (see Chapter 1), could never have been brought about through the mere chance accumulation of similar variations in each of many successive generations. In practice, a pigeon fancier is perhaps struck by one particular pigeon having a slightly shorter beak than others, while another fancier is struck by a pigeon having a rather longer beak; on the acknowledged principle that "fanciers do not and will not admire an average standard, but like extremes," they both go on choosing and breeding from birds with longer and longer beaks, or from birds with shorter and shorter beaks, generation after generation after generation. This has certainly happened with the sub-breeds of the tumbler pigeon.

Similarly, we may suppose that at an early period of history, the men of one nation or district required swifter horses, while those of some other place required stronger and bulkier horses. The horses would not look very different from each other early in the process. But over time, from the continued selection of the fastest horses in the one case, and of stronger ones in the other case, the differences would be greater, and greater, and greater, and eventually we would end up with two subbreeds. Ultimately, after the lapse of centuries, these subbreeds would become converted into two well-established and distinct breeds. As the differences became greater, the inferior animals with intermediate traits, being neither especially swift nor very strong, would not have been used for breeding and thus will have tended to disappear from the population. Here, then, we see in man's productions the action of what may be called "the principle of divergence" causing differences—at first barely detectable—to steadily increase with

(A)

(B)

Figure 4.10 (A) Racehorse; bay stallion. (B) Cart horse; white Shire draft horse.

each generation, and the breeds to then diverge more and more in character over time, both from each other and from their common parent.

But how, it may be asked, can such a principle apply in nature, without the involvement of selection by humans? I believe that it does indeed apply, and does so most efficiently, although it was a long time before I saw how. It follows from this simple circumstance: the more the descendants of any one species become diversified in structure, physiology, and habits, the more will they also be better enabled to seize on niches and habitats not previously occupied by other members of that species, and so be enabled to increase in number.

We can clearly see this in the case of animals with simple habits. Take the case of some carnivorous quadruped whose population size has long ago reached the maximum that can be supported in some particular place. Unless there is some change in the climate or some other conditions in the area, then if its natural power of increase is allowed to act, the animals can increase in numbers only if their varying descendants are able to seize on places presently occupied by other animals: some of these descendants being enabled to feed on new kinds of prey, either dead or alive, for example; some inhabiting different lifestyles than their parents—climbing trees or frequenting water, or even becoming less carnivorous and feeding more on plant material. The more diversified in habits and structure the descendants of our carnivorous animals become, the more new niches they will be enabled to occupy.

What applies to one animal will apply throughout all time to all animals, as long as they vary in individual characteristics—for otherwise natural selection cannot act, and can accomplish nothing.

The same arguments apply to plants. Experiments show that if a plot of ground is sown with just one species of grass, and a similar plot is sown with seeds of several grass species belonging to a number of distinctly different genera, a greater number of plants and a greater total weight of dry plant material can be raised in the diversified plot than in the single-species plot. Imagine, then, that we have a species of grass that continues to vary and that there is a gradual selection for those varieties that come to differ from each other—even to only a slight degree—in the same manner as do the members of distinct species and genera. Over time then, a greater number of individual plants of this species, including its modified descendants, would succeed in living on the same piece of ground. And we know that each species and each variety of grass is every year releasing almost countless numbers of seeds, and is thus striving to the utmost, in a sense, to increase in number. Consequently, in the course of many thousands of generations, the most distinct varieties of any one species of grass would have the best chance of surviving, reproducing, and increasing in numbers, and thus of overwhelming and replacing the less distinct varieties of that same species. And once varieties become very distinct from each other, they eventually earn recognition as separate species.

The truth of this basic principle—that the greatest amount of life in any area can be supported by great diversification of structure and function—is seen under many natural circumstances. In an extremely small area, especially if it is freely open to immigration, the contest between individuals must be very severe; in such situations we always find great diversity in its inhabitants. For instance, I found that a small piece of turf, only 3 feet wide and 4 feet long, that had been exposed to the same conditions year after year for many years, supported 20 species of plants belonging to 18 different genera and eight different orders, which shows just how much these plants differed from each other. And so it is with the plants and insects on small and uniform islands, or in small freshwater ponds. Similarly, farmers find that they can raise the most food on any given area of land by periodically sowing the seeds of plants belonging to the most different orders; nature similarly follows what may be called a simultaneous crop rotation. Most of the animals and plants that live together on any small piece of suitable ground could live on it, and may be said to be striving to the utmost to live there; but where they come into the closest competition, the advantages of diversification of structure, with the accompanying differences of habit and constitution, determine that the inhabitants that jostle each other the most closely shall, as a general rule, show great variation in how they live and thus belong to what we call different genera and different orders.

The same principle is seen in the naturalization of plants through human intervention in foreign lands. You might think that the invasive plants that have become successfully established in any new land would generally be species that are closely related to the indigenous, native species, particularly if the native species had been specially created for life in its own country. Wouldn't successful invaders then be expected to share similar traits? We might also have expected that successfully introduced non-native plants would belong to a few groups more especially adapted to certain conditions in their new homes. But the actual case is quite different; as the Swiss botanist Alphonse de Candolle has well remarked in his great and admirable work, successful invasions increase diversity in an area in proportion to the number of the native genera and species, and far more *in new genera* than in new species; thus plant diversity in the area is increased greatly through these invasions. Differing characteristics permit long-term coexistence.

To give but one example, the most recent edition of Dr. Gray's *A Manual of the Botany of the Northern United States* (1856), lists 260 naturalized plant species[8] belonging to 162 different genera. These non-native plants differ considerably from the indigenous species: at least 100 of the 162 successfully invading genera are not represented at all among the native species in that area, thus greatly increasing the number of plant genera now living in the United States, and increasing plant diversity enormously.

---

[8] Naturalized species are non-native plants: exotic plants that escaped from cultivation after being deliberately introduced to an area, and that are now living on their own in their new habitat.

Considering the nature of the introduced or invasive plants and animals that have struggled successfully with the native species in any area and become "naturalized" (i.e., become permanent members of those communities) may help us to understand how some of the native species would now have to become modified if they were to gain an advantage over their new compatriots. We may at least infer that substantial diversification of structure, at the level of generic differences, would benefit them.

Diversification of structure among the inhabitants of the same region is advantageous in the same way that physiological division of labor among the various organs of the body is advantageous—a subject so well elucidated by that expert on the biology of lower animals Henri Milne-Edwards. No physiologist doubts that if an animal's stomach is specifically adapted for digesting vegetable matter, the organism possessing that stomach will draw most of its nutrition from that material. On the other hand, a stomach that is specifically adapted for obtaining nutrients from flesh alone will draw most of its nutrients from those materials. Thus in the general economy of any land, the more widely and perfectly the animals and plants of that area are specialized for different habits of life, the greater the total number of individuals that will be able to support themselves there, by avoiding competition. A set of organisms with their organization only slightly diversified could hardly compete successfully for long with a set of organisms that were more perfectly diversified in structure. For example, the Australian marsupials are divided into groups differing fairly little from each other, and represent our carnivorous, ruminant, and rodent mammals only feebly; we can surely doubt whether those marsupials would be able to compete successfully with members of these well-developed orders if they were introduced to the United Kingdom, or vice versa. In the Australian mammals we can see that the process of diversification is still an early and incomplete stage of development.

Effects of Natural Selection on the Descendants of a Common Ancestor, Through Divergence of Character and Extinction

After the previous, unfortunately brief discussion, we may assume that the modified descendants of any one species will succeed so much better as they become more diversified in structure (and function), and thus become enabled to encroach upon the habitats and lifestyles presently occupied by other beings. Now let us see how this basic principle of benefitting from a divergence in character tends to act in combination with the principles of natural selection and extinction.

Figure 4.11 helps us to visualize this rather perplexing subject. The capital letters A through L[9] along the x-axis represent the 11 species of some

---

[9] Darwin skipped the letter J in his diagram of 11 species, so we have skipped it as well!

single genus that is widely represented in its own country. These 11 species differ from each other to different extents, as is so often the case in nature, and so I have placed them at different distances away from each other on the diagram. The distance between species A and B, for example, is much smaller than that between species D and E, indicating that species A and B are more similar to each other than are species D and E. Species A and D are obviously even more different from each other. I have said that all of these species belong to a large genus because as we saw in Chapter 2, more species, on average, vary in large genera than in genera with relatively few species. Moreover, the varying species of large genera tend to show more varieties. We have also seen that the most common and the most widely distributed species vary more than do the rare species with restricted ranges.

So let species A be a common, widely-diffused, and varying species belonging to a genus that is large in its own country. The branching and diverging dotted lines of unequal lengths that extend from species A near the bottom of the figure represent its varying offspring. The variations are extremely slight, but very diversified; they do not all appear simultaneously, but often after long intervals of time, nor do they all endure for the same

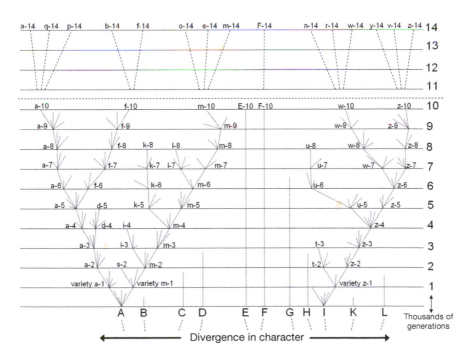

Figure 4.11 A schematic illustration of how morphological (and physiological) diversity increases over time. Organisms close to each other along the x-axis resemble each other more than do organisms further apart. The y-axis represents time, with time increasing from the bottom of the graph upwards, representing thousands of generations or even more.

amounts of time. Only those variations that are in some way profitable to their possessor will be preserved or naturally selected, because those are the variations that will minimize competition with the "parents" by allowing those offspring to exploit new resources. If a dotted line reaches one of the horizontal lines, and is marked there by a lowercase numbered letter, a sufficient amount of variation is supposed to have been accumulated by then to form it into a fairly well-marked variety—something that would be thought worth recording in a scientific publication.

Suppose that the distance between each horizontal line in the diagram represents 1,000 generations. After 1,000 generations, then, species A is shown to have produced two fairly well-marked varieties, namely varieties a-1 and m-1. Since the tendency to vary is in itself hereditary, these varieties will also tend to vary, and commonly in nearly the same manner as did their parents. Moreover, these two varieties, being only slightly modified forms, will tend to inherit those advantages that made their parent (species A) more successful and numerous than most of the other inhabitants in the same region, and will also partake of those more general advantages that made the genus to which the parent species belonged a large genus in its own region. All these circumstances favor the production of new varieties.

If, then, these two varieties continue to vary, the most divergent of their variations will generally be preserved during the next thousand generations. And after this interval, you see that variety a-1 in our diagram has now produced variety a-2, which will, owing to the principle of divergence, differ more from ancestor A than did variety a-1. Variety m-1 in the diagram is seen to have produced two new varieties, namely s-2 and m-2, which will differ not only from each other but even more so from their common parent, ancestor A. We may continue this process by similar steps for any length of time, with some of the varieties after each 1,000 generations producing only a single variety (e.g., a-6), but in a more and more modified condition; with some producing two or three varieties (e.g., the decendants of m-1); and some failing to produce any new varieties at all (e.g., species D and G). Thus the varieties or modified descendants of the common parent, ancestor A, will generally go on increasing in number and diverging in character. Figure 4.13 summarizes this process to the ten-thousandth generation, and under a condensed and simplified form beyond that up to the fourteen-thousandth generation at the top of the figure.

I do not suppose that the actual process ever goes on as regularly as is shown in the diagram, nor that it goes on continuously. It is far more likely that each form remains unaltered for long periods of time and then again undergoes modification for a time. Nor do I suppose that it is always the most divergent varieties that are preserved: a medium form may well endure for a long time, and may or may not produce more than one modified descendant. Natural selection will always act according to the nature of the places that are either unoccupied or not perfectly occupied by other beings, and this will depend on infinitely complex relations and interactions. But as a general rule, the more that the descendants of any one species can become increasingly diversified in structure, the more places they will be enabled to succeed in and the more their modified progeny will increase in numbers. In our diagram, the line of succession is broken at regular intervals by small numbered letters marking the successive forms that have become sufficiently distinct to be recorded as varieties. But these breaks are imaginary: I might have inserted them anywhere, after intervals long enough to allow a considerable amount of divergent variation to accumulate.

All the modified descendants from a common and widely distributed species belonging to a large genus will tend to benefit from the same advantages that made their parents successful in life; thus they will generally go on multiplying in number as well as diverging in character. This is shown in our figure by the several diverging branches proceeding from species A. The modified offspring from the later and more highly improved branches in the lines of descent will often take the place of the earlier and less improved branches and so destroy them: this is represented in the diagram by some of the lower branches simply ending before they reach the upper horizontal lines. In some cases the process of modification will probably be confined to a single line of descent and the number of modified descendants will not be increased, even though the amount of divergent modification may have been great. Such a case would be represented in the diagram if all the lines proceeding from ancestor species A were removed, except for the one extending from a-1 to a-10. In the same way, the English racehorse and the English pointer dog (Figure 4.12) have both apparently gone on slowly diverging in character from their ancestral stocks, without either one having given off any fresh branches or races.

After 10,000 generations in our diagram, species A has produced three forms (a-10, f-10, and m-10) which, from having diverged in character during the succeeding 100s of generations, will have come to differ largely, although perhaps unequally, both from each other and from their common parent. If we suppose the amount of change between each horizontal line in our diagram to be extremely small, these three forms may still be only well-marked varieties, so that we would now have, for example, variety a-10 and variety m-10. But suppose that the steps in the process of modification had been more numerous or greater in degree; in that case, these three forms will likely have been converted into either doubtful/questionable species or perhaps even into well-defined species; i.e., a-10 and m-10 might now be distinct species. **Thus the diagram illustrates the steps by which the small differences distinguishing varieties**

Figure 4.12 English pointer.

can slowly increase to become the larger differences distinguishing species.
By continuing the same process for an even greater number of generations
(as shown at the top of the diagram in a simplified manner), we get eight
species, marked by the letters between a-14 and m-14, all descended—over
many 1000s of generations—from species A. This, I believe, is how species
are multiplied and genera are formed over time.

In a large genus more than one species would probably vary. I have
assumed in Figure 4.11 that after 10,000 generations a second species (species I) has produced, by analogous steps, either two well-marked varieties
(w-10 and z-10) or two species, according to the amount of change that
is represented between the horizontal lines. After 14,000 generations, six
new species, marked by the letters n-14 to z-14 have been produced. In
any genus, the species that already differ greatly in character from each
other will tend to produce the greatest number of modified descendants;
for these will have the best chance of seizing on new and widely different
niches. In the diagram I have therefore chosen the extreme species (A)
and the nearly extreme species (I) as those that have varied the most and
have given rise to new varieties and species. The other nine species in
our genus, marked by the other capital letters, may continue to transmit
unaltered descendants for long but unequal periods of time; this is shown
in the diagram by the dotted lines unequally prolonged upwards (species
D, E, F, and G, for example).

But during the process of modification that is represented in the diagram, another of our principles, namely that of extinction, will also have
played an important part. In any stable area that is fully stocked with species, natural selection acts by the selected form having some advantage in
the struggle for life over other forms. Thus there will be a constant tendency
for the improved descendants of any one species to supplant and exterminate their predecessors in each stage of descent, including the form that
first gave rise to them. Recall that competition will generally be most severe
between those forms that are most nearly related to each other in habits,
physiology, and structure. Thus all the intermediate forms between the earlier and later stages—that is between the less improved and more improved
states of the same species—as well as the original parent species itself, will
generally tend to eventually become extinct. So it probably will be with
many whole collateral lines of descent, which will eventually be conquered
by later and improved lines. On the other hand, the modified offspring of a
species might end up living in a different area, or become quickly adapted
to some quite new niche in the original area, so that offspring and parents
do not come into competition; if so, then both may continue to coexist.

If we now assume that our diagram represents a considerable amount
of modification, the original species A and all the earlier varieties will have
become extinct, having been replaced by eight new species: a-14 to m-14.
Similarly, species I is now extinct, but it has been replaced by six new species, n-14 through z-14.

But we may go still further. The various original species in our genus were supposed to resemble each other to different degrees, as is so generally the case in nature: species A is more closely related to species B, C, and D than to the other species in our diagram; and species I is more closely related to species G, H, K, and L than to the others. Species A and species I were also supposed to be very common and widely distributed species, so that they must originally have had some advantage over most of the other species of the genus. Their modified descendants, 14 by the fourteen-thousandth generation in the hypothetical scenario of Figure 4.11, will probably have inherited some of these same advantages: they have also been modified and improved in a diversified manner at each stage of descent, so as to have become adapted to many related places in the natural economy of their country. It seems extremely probable, then, that they will have taken the places of—and exterminated—not only their parents A and I, but also some of the original species that were most nearly related to and most like their parents. Thus, very few of the original species will have been successful in transmitting offspring to the fourteen-thousandth generation. In our diagram, only one (species F) of the two species (species E and F) that were least closely related to the other nine original species has transmitted descendants to this late stage of descent.

In our diagram, 15 species have now descended from the original 11 species. Owing to the divergent tendency of natural selection, the extreme amount of difference in character between species a-14 and z-14 will be much greater than that between even the most distinct of the original 11 species. The new species, moreover, will be allied to each other in a widely different manner. Of the eight descendants from species A, the three marked a-14, q-14, and p-14 will be closely related, since they have all recently branched off from their common ancestor, a-10; b-14 and F-14 will also be closely related, from their having diverged longer ago from a-5, but will be in some degree distinct from the three first-named species; and lastly, although species o-14, e-14, and m-14 will also be closely related to each other, they will differ greatly from the other five species, having diverged at the very beginning of the process of modification. Indeed, species o-14, e-14, and m-14 may now constitute a separate subgenus or even a distinct genus.

The six descendants from species I will form two subgenera or genera. But as the original species I differed so largely from species A, standing nearly at the extreme end of the original genus on our chart, the six descendants from ancestor I will, owing to inheritance alone, differ considerably from the eight descendants of ancestor A. Moreover, the two groups are shown in the figure to have gone on diverging in different directions. The species that originally had characteristics that were intermediate between those of species A and species I have all gone extinct, except for species F, and left no descendants. This is a very important point. Thus the six new species descended from ancestor I, and the eight descendants from ancestor

A, will now have to be ranked as being in very distinct genera, or even in distinct subfamilies.

This is, I believe, how two or more genera are eventually produced by descent with modification over many, many generations from two or more species of the same genus. And the two or more parent species are presumably descended from some one species of an earlier genus. In our diagram, this is indicated by the broken lines beneath the capital letters, converging in subbranches downwards towards a single point; this point (e.g., the point formed by following the lines down from species A, B, C, and D) represents what is presumably the ancestral species that eventually gave rise to our several new subgenera and genera.

It is worth reflecting for a moment on the character of the new species F-14, which has not diverged much in character and has retained the form of the original species, species F, either unaltered or altered only slightly. In this case, its affinities to the other 14 new species will be of a curious and circuitous nature. Being descended from a form that stood between the parent species A and I, now supposed to be extinct and unknown, it will now in some degree have characteristics that are intermediate between those of the two groups descended from these two species. But as these two groups have gone on diverging in character from the prototype of their parents, the new species (F14) will not be directly intermediate between the two, but rather between *types* of the two groups; every naturalist will be able to call such actual cases to mind.

In Figure 4.11, I suggested that each horizontal line represented 1,000 generations. But each line may instead represent a million generations, or even more. Each line may also represent a section of the successive layers of the earth's crust including remains of organisms that are now extinct. We will refer again to this subject when we come to our chapter on geology (Chapter 11), and I think that we will then see that this diagram also throws considerable light on the relationships among extinct organisms, which though generally belonging to the same orders, families, or genera represented by those now living, yet are often at least somewhat intermediate in character between the members of existing groups. This makes good sense when we consider that the extinct species lived at various remote epochs when the branching lines of descent had diverged less from each other than they do now.

I see no reason to limit the process of modification that I have just explained to the formation of genera alone. If, referring again to Figure 4.11, we suppose the amount of change represented by each successive group of diverging dotted lines to be very great, then those forms marked a-14 to p-14, those marked b-14 and f-14, and those marked o-14 to m-14, will form three very distinct genera. We shall also have two very distinct genera descended from species I that differ greatly from the descendants of species A. These two groups of genera will thus form two distinct families—or even orders—of organisms, depending on the amount of divergent modification that is represented in the diagram. And the two new families (or orders) are descended from two species (species A and species I) of the original genus, and these are, in turn, supposed to be descended from some still more ancient and unknown form.

We have seen that in each country it is the species belonging to the larger genera, containing many species, which most often present varieties (i.e., incipient species). This indeed might have been expected: as natural selection acts through one form having some advantage over other forms in the struggle for existence, it will chiefly act on those that already have some advantage over others; and the largeness of any group shows that the species within that group have inherited from a common ancestor some advantages in common. Thus the struggle for the production of new and modified descendants will mainly lie between the larger groups that are all trying to increase in size. One large group will slowly conquer another large group and reduce its numbers, thus lessening the chance of further variation and improvement among members of the conquered group. Within the same large group, the later and more highly perfected subgroups (e.g., subfamilies, or subspecies) will, from branching out and seizing on many new niches, constantly tend to supplant and destroy the earlier and less improved subgroups; small and broken groups and subgroups will then finally disappear. Looking to the future, we can predict that the groups of organisms that are now large and triumphant, and which are least broken up—that is, those that have as yet suffered the least extinction among members—will continue to increase for a long time.

But which groups will ultimately prevail, no one can predict: we know with certainty that many groups that were formerly most extensively developed have now become extinct. Looking still further out into the future, we may predict that, owing to the continued and steady increase in size of the larger groups, a multitude of smaller groups will eventually become utterly extinct and leave no modified descendants. Consequently, of the species living at any one period, extremely few will successfully transmit descendants to a remote and distant future.

According to this view, very few of the more ancient species have transmitted descendants to the present day, and, as all the descendants of the same species form a taxonomic class, we can understand how it is that so few classes now exist in each main division of the animal and plant kingdoms. Although few of the most ancient species may have left modified descendants to the present day, yet at even the most remote geological period, the Earth may have been almost as well populated with species of as many genera, families, orders, and classes as we have now. I shall have more to say about this subject later, in the chapter on classification (Chapter 14).

On the Degree to Which Organisms Tend to Advance in Complexity

Natural selection acts exclusively by preserving and accumulating variations that are beneficial under the ecological and physical conditions to which each creature is exposed at all periods of its life. The ultimate result is that each creature tends to become better and better adapted to the conditions that surround it. This improvement inevitably leads to a gradual advancement in organization for most organisms throughout the world. But here we enter on a very intricate subject, for naturalists have not yet defined to each other's satisfaction what they mean by an "advance in organization." Among the vertebrates, both the degree of intelligence and the degree to which the body structure approaches that of humans are clearly relevant to the argument. It might be thought that the amount of change that the various parts and organs pass through in their development from the embryo to maturity could serve as a valid standard of comparison; but there are cases, as with certain parasitic crustaceans,[10] in which several parts of the structure have become less well-developed, so that the mature animal cannot be considered—in terms of morphology—more advanced than its larval stage.

Development in such parasitic organisms thus involves the loss of complexity and a move toward simplification rather than what we normally think of as an "advance." The embryologist Karl Ernst Baer's standard seems the most widely applicable and the best. In his view, "advanced" refers to the amount of differentiation of the parts of the same organism—in the adult state, I should be inclined to add—and their specialization for different functions. Alternatively, we might adopt Monsieur Milne-Edwards' view of advancement as the completeness of the division of physiological labor.

But we shall see how obscure this subject is if we look, for instance, to fishes. Some naturalists rank those which, like the sharks, are most like amphibians as the most advanced, while other naturalists rank the common bony fishes as the most advanced, inasmuch as they are more strictly "fish-like" and differ most from members of the other vertebrate classes. We see the obscurity of the subject still more plainly by turning to plants, amongst which a standard of intelligence is of course quite excluded. Some botanists consider the highest-level plants to be those having every organ—such as sepals, petals, stamens, and pistils—fully developed in each flower, whereas other botanists, probably with more truth, consider those plants having their various organs much modified and reduced in number to be the most advanced.

If we consider the amount of differentiation and specialization of the several organs in each adult (including the advancement of the brain for intellectual purposes) in evaluating the standard of organization, then natural selection clearly leads toward that standard. All physiologists admit that it is an advantage for organisms to have more specialized organs, inasmuch as in this state those organs will perform their functions better. Thus the accumulation of variations tending toward increasing specialization is within the power of natural selection. On the other hand, bearing in mind that all

---

[10] The Crustacea is a group of arthropods containing such animals as crabs, shrimp, barnacles, and copepods.

organisms are striving to increase their numbers at a high rate and to seize on every unoccupied or less well-occupied place in the economy of nature, we can see that it is quite possible for natural selection to gradually fit an organism to a situation in which several organs would be superfluous or useless: in such cases there would be movement toward reduced complexity. Whether organization in general has actually advanced from the remotest geological periods to the present day will be more conveniently discussed later, in the chapter on geological succession (Chapter 11).

But one may object that if all organic beings thus tend to become more complex as their species develops, then how is it that throughout the world a multitude of the lowest forms still exist? And how is it that within each great class of organisms, some forms are far more highly developed than others? Why haven't the more highly developed forms everywhere supplanted and exterminated the less complex? The French naturalist Jean-Baptiste de Monet Lamarck, who believed in an innate and inevitable tendency toward perfection in all organisms, seems to have felt this difficulty so strongly that he was led to suppose that new and simple forms are continually being produced by spontaneous generation, something that science has yet to confirm. However, the continued existence of lowly organisms poses no difficulty for us: natural selection, or the survival of the fittest, does not necessarily require progressive development. It only takes advantage of such variations as arise and are beneficial to each creature under its complex relations of life. And it may be asked, what advantage would it be to a paramecium (Figure 4.13) or to some other protozoan, or to an intestinal worm, or even to an earthworm, to be more highly organized? If it were not advantageous, then those forms would be left unimproved or but little improved by natural selection, and might remain for indefinite ages in their present lowly condition. And geology tells us that some of the lowest forms, such as the ciliated protozoans and amoebae, have remained in nearly their present state for an enormous period of time. But to suppose that most of the many now-existing lowly forms have not advanced in the least since

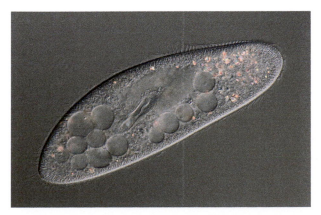

Figure 4.13 *Paramecium caudatum.*

Figure 4.14 A large grey reef shark (*Carcharhinus amblyrhynchos*).

the first dawn of life would be extremely rash; indeed, every naturalist who has dissected some of these so-called lowly beings must have been struck with how wondrous and beautifully organized they were.

Nearly the same remarks apply if we look to the different grades of organization within any one great group. Within the Vertebrata, for example, how can we explain the co-existence of mammals and fish? Amongst mammals, how can we explain the coexistence of humans and the platypus? Amongst fishes, how can we explain the coexistence of the shark (Figure 4.14) and the lancelet (*Amphioxus*) (Figure 4.15), whose extremely simple structure approaches that of the invertebrates? But mammals and fishes hardly ever come into competition with each other; thus the advancement of the entire class of mammals—or of certain members in this class—to the highest grade would not lead to their taking the place of fishes. Physiologists believe that the brain must be bathed by warm blood to be highly active, and this requires aerial respiration. Thus, warm-blooded aquatic mammals are disadvantaged by having to come continually to the surface to breathe. With fishes, members of the shark family would not tend to supplant the lancelet; according to the German biologist Fritz Müller, the lancelet has an anomalous annelid (the phylum of segmented worms that includes marine worms, earthworms and leeches) as its only companion and competitor, on the barren sandy shores of

Figure 4.15 A Lancelet (*Amphioxus lanceolatus*). Lancelets of various species are found all over the world in shallow temperature and tropical seas, usually half-buried in sand. They are used as a food source in Asia, and are commonly studied for what they may tell us about the evolution of vertebrates.

South Brazil where it lives. The three most primitive orders of mammals—the marsupials, the edentata (which includes the armadillos, anteaters, and aardvarks),[11] and the rodents—coexist in South America in the same region with numerous monkeys, and they also probably don't interact with each other very much. Although organization, on the whole, may have advanced and may still be advancing throughout the world, yet the scale will always present many degrees of perfection; the high advancement of certain whole classes, or of certain members of each class, does not at all necessarily lead to the extinction of groups that they don't directly compete against. In some cases, as we shall see later, lowly organized forms appear to have survived or even flourished to the present day simply from inhabiting confined or peculiar habitats where they have been subjected to less severe competition, and where the relatively small numbers of individuals have reduced the chances of favorable variations arising.

Finally, I believe that many lowly organized forms now exist throughout the world from a variety of causes. In some cases, appropriate variations or individual differences of a favorable sort may never have arisen for natural selection to have acted on and accumulated. In no case, probably, has there been enough time to permit the utmost possible amount of development. In some few cases there has been what we must call simplification of organization. But the main cause lies in the fact that under very simple conditions of life, a high organization would be of no value—indeed, it might even be a disadvantage, as being of a more delicate nature, and more liable to be put out of order and injured.

Looking to the dawn of life, when all organisms, we may assume, presented the simplest of structures, how, it has been asked, could the first steps in the advancement or differentiation of parts have arisen? As we have no facts to guide us, speculation on the subject is almost useless. It is, however, an error to suppose that there would be no struggle for existence, and, consequently, no natural selection, until many forms had been produced: variations in a single species inhabiting an isolated region might be beneficial, and thus the whole mass of individuals might be modified, or two distinct forms might arise. But, as I remarked toward the end of the introduction, no one ought to be surprised that there is much that is still unexplained about the origin of species, particularly if we allow for our profound ignorance of the mutual interrelationships and interactions among the inhabitants of the world at the present time, and still more so during past ages.

Convergence of Character

Mr. Hewett Cottrell Watson, the English botanist I mentioned in Chapter 2, thinks that I have overrated the importance of divergence of character

---

[11] These animals have subsequently been distributed among three new orders, including the Xenarthra.

(although he does, apparently, believe in it), and that convergence, as it may be called, has likewise played a part. If two species belonging to two distinct though related genera had both produced a large number of new and divergent forms, it is conceivable that these forms might resemble each other so closely that they would all end up being classified under the same genus; in this case, then, the descendants of two distinct genera would converge into a single genus. But it would in most cases be extremely rash to explain such a close and general similarity of structure in the modified descendants of widely distinct forms through convergence. The shape of a crystal is determined solely by the molecular forces acting within it and it is not surprising that dissimilar substances should sometimes assume the same form. But with living organisms, the form of each depends on an infinitude of complex relations, namely 1) on the variations that have arisen; 2) on the nature of the variations that have been preserved or selected, which depends of course in part on the surrounding physical conditions, and even more so on the surrounding organisms with which each being has come into competition; and 3) on inheritance from innumerable ancestors, all of which have had their forms determined through equally complex relations. It is incredible to think that the descendants of two organisms that were originally very different from each other should ever afterward converge so closely as to make those descendants now nearly identical. If this had occurred in the past, then we should now encounter the same form, independently of genetic connection, recurring in widely separated geological formations; the balance of evidence argues again any such admission.

Mr. Watson has also objected that the combination of natural selection and divergence of character would tend to create an indefinite number of species. Considering only physical conditions, it seems likely that a sufficient number of species would soon become adapted to all considerable diversities of heat, moisture, and so forth, and that no new species would be formed. But I fully believe that the ecological interactions among organisms are more important; and as the number of species in any land goes on increasing, the ecological interactions among them must become more and more complex over time. Consequently, at first sight it does seem that there must be no limit to the amount of profitable diversification of structure that might be produced, and thus no limit to the number of species that might be created. We don't know whether or not even the most prolific areas are now fully stocked with species; at the Cape of Good Hope (located at the southernmost tip of Africa) and in Australia, places that now support such an astonishing number of species, many European plants have recently become part of the natural flora, showing quite clearly that not all available niches were previously filled. But geology shows us that over the past 60 million years or so the number of shelled animal species (e.g., snails, clams, and brachiopods) (Figure 4.16) has not increased greatly or at all, and the same is true of the number of mammal species over the past 30 million years or so. What then prevents an indefinite increase in the number of species living at any given time?

Figure 4.16 A dried specimen of a brachiopod (*Terebratulina septentrionalis*). The animal's lophophore, which generates feeding currents and collects food particles, is visible inside the shell.

There must be some limit to the amount of life that can be supported in any area, depending as it does on physical conditions. Thus, if an area is inhabited by very many species, each or nearly each species will be represented by only a few individuals. Such species will then be liable to extermination from occasionally severe fluctuations in climate or in the numbers of their enemies. The process of extermination in such cases would be rapid, whereas the production of new species must always be slow. Imagine the extreme case of having only one individual representing each of many species in England; the first unusually severe winter or the first unusually dry summer would exterminate thousands upon thousands of species. Rare species—and every species will become rare if the total number of species in any country becomes indefinitely increased—will, as I have explained earlier, present within any given period very few beneficial variations. Consequently, the process of giving birth to new species would be slowed. Moreover, when any species becomes very rare, close inbreeding will help to exterminate it; a variety of authors have suggested that such inbreeding accounts for the deterioration, for example, of the aurochs (the large ancestors of our domestic cattle) in Lithuania, of red deer in Scotland, and of bears in Norway. Lastly, and perhaps most importantly, a dominant species that has already beaten many competitors in its own home will tend to spread and supplant many others. Alphonse de Candolle has shown that those species that spread widely tend to spread *very* widely indeed. Thus they will tend to supplant and exterminate several species in several areas, thereby preventing an endless increase in new species throughout the world. How much weight to attribute to these several considerations I will not pretend to know. But taken together they must limit, in any given region, the tendency to an infinite and endless increase in the numbers of species alive at any given time.

Summary

The following points cannot be disputed: 1) Under changing conditions of life, living organisms present individual differences in almost every part of their structure; 2) Owing to the exponential rate at which individuals in a population increase, there will inevitably be a severe struggle for life at some age, or in some season, or in some year. It follows then, that a great diversity in structure, physiology, and habits is highly advantageous within a species, considering the infinite complexity of the relations of all organisms to each other. Thus it would be most extraordinary if no variations had ever been useful to each being's own welfare, in the same way that so many variations have proved useful to us.

But if variations useful to any organism do occur, then individuals possessing those variations will have the best chance of surviving in the struggle for life and reproducing. From the strong principle of inheritance, these individuals will tend to produce offspring with similar characteristics. I have called this principle "natural selection." It leads inevitably to the improvement of each creature in relationship to the ecological and physical conditions of its life, and, consequently, in most cases, to what must be regarded as an advance in organization. Nevertheless, low and simple forms will long endure if they are well suited to their simple conditions of life.

Because traits tend to be inherited at the same ages that they were first developed in the parental generation, natural selection can select for advantageous traits in eggs, seeds, larvae, or juveniles as easily as in the adult. Among many animals, sexual selection will have aided ordinary selection by giving the greatest number of offspring to the most vigorous and best-adapted males. Sexual selection will also encourage the development of characteristics (i.e., traits) useful to the males alone in their struggles or rivalry with other males. These traits will in turn be transmitted to one sex or to both sexes, according to the form of inheritance that prevails in that species.

Whether or not natural selection has really acted in this way in adapting the various forms of life to their particular environments and lifestyles must be judged by the weight of evidence provided in the following chapters. But we have already seen how it involves extinction; indeed, the large role played by extinction is very clearly seen in looking at the geological record. Natural selection also leads to divergence of character; the more that organisms differ in structure, behavior, and physiological traits, the larger the total number of organisms that can be supported in any area, through reduced competition. We see clear proof of this in looking at the inhabitants of any small piece of land, and in the species that have become naturalized in foreign lands. Thus, as the descendants of any one species are modified over time, and during the incessant struggle of all species to increase their numbers, the more diversified the descendants become, the more likely they are to succeed in the constant battle for life. Thus the small differences distinguishing varieties of the same species tend to steadily increase until they eventually become large enough to form distinct species, or even a new genus.

We have seen that it is the common, widely diffused, and wide-ranging species belonging to the larger genera within each class that vary the most; these tend to transmit to their modified offspring that superiority that now

makes them dominant in their own habitats. Because natural selection leads
to divergence of character and to much extinction of the less improved and
intermediate forms of life, the nature of the affinities as well as the generally sharply defined distinctions between the innumerable organisms in
each class throughout the world may be explained. It is a truly wonderful
fact—the wonder of which we are apt to overlook from familiarity—that
all animals and all plants throughout all time and space should be related
to each other in groups, and in a very particular way: varieties of the same
species are always most closely related; species of the same genus are still
related, but less closely and to unequal degrees, forming sections and sub-
genera; species belonging to different genera are much less closely related;
and genera are related to each other to different degrees, forming subfamilies, families, orders, subclasses, and classes (see Figure 1.5). The several
subordinate groups within any class of organisms cannot be ranked in a
single file, but seem instead to be clustered around points, and these are
clustered around other points, and so on in almost endless cycles. If species
had been independently created, we would not be able to explain this kind
of arrangement; but it is explained very nicely through inheritance and the
complex actions of natural selection, involving extinction and divergence
of characteristics, as we have seen illustrated in Figure 4.11.

The relationships of all members of the same class have sometimes
been represented by a great tree. I believe that this simile largely speaks the
truth, with the green and budding twigs representing the existing species,
and those produced during former years representing the long succession
of extinct species. At each period of growth, all of the growing twigs have
tried to branch out on all sides, and to smother and kill the surrounding
twigs and branches, in the same manner as species and groups of species
have at all times overwhelmed other species in the great battle for life. The
limbs that we now see divided into great branches, and these into lesser and
lesser branches, were themselves once budding twigs when the tree was
young, and this connection of the former and present buds by ramifying
branches may well represent the classification of extinct and living species
in groups within groups. Of the many twigs that flourished when the tree
was a mere bush, only two or three, now grown into great branches, still
survive and bear the other branches. So it is with the species that lived
during ancient geological periods: very few have left living and modified
descendants. From the first growth of the tree, many a limb and branch
has decayed and dropped off; those fallen branches of various sizes now
represent whole orders, families, and genera that no longer have living
representatives, and which are known to us only as fossils. As we here
and there see a thin straggling branch springing from a fork low down in
a tree, and which by some chance has been favored and is still alive on its
summit, so we occasionally see an animal like the platypus or the lungfish,
which in some small degree connects two large branches of life, and which
has apparently been saved from fatal competition with other organisms by

having inhabited an unusual niche with limited competition and assault from enemies. As buds give rise by growth to fresh buds, and as these, if vigorous, branch out and overgrow on all sides many a feebler branch, so I believe it has been with the great Tree of Life, which fills with its dead and broken branches the crust of the earth, and covers the surface of our living planet with its ever-branching and beautiful ramifications.

Key Issues to Talk and Write About

1. This chapter is about natural selection. As Darwin describes it, how is natural selection similar to the kind of selection (artificial selection) that people have used in creating our domesticated animals and agricultural crops? How do the two forms of selection differ?

2. Find out two interesting things about one of the people that Darwin mentions in this chapter. Choose from the following:

Jean-Henri Fabre	Christian Konrad Sprengel
Asa Gray	Henri Milne-Edwards
Thomas Henry Huxley	Karl Ernst Baer
Joseph Hooker	Jean-Baptiste de Monet Lamarck
Joseph Gottlieb Kölreuter	Alphonse de Candolle
Hewett Cottrell Watson	Fritz Müller
Moritz Friedrich Wagner	

3. Carefully read the paragraph that begins "Physical peculiarities often appear either in males or females..." (see page 70). For the chosen paragraph, what are the two or three main points that Darwin wishes to get across?

   Now summarize those points in a single sentence: your sentence should include all the major points, be accurate, make sense to someone who has not read the original paragraph, and be in your own words.

4. Following the instructions given near the end of Chapter 1 (see page 28), write a one-sentence summary of that paragraph, being careful to include all the key points that you think Darwin is trying to get across. Try the same exercise with the paragraph that begins, "A structure used only once..." (see page 69).

5. Rewrite the following sentence from Darwin's original, to make it more concise and clear:
   Natural selection acts exclusively by the preservation and accumulation of variations, which are beneficial under the organic and inorganic conditions to which each creature is exposed at all periods of life.

6. How does Darwin explain the continued existence of the platypus, and other "living fossils"?

7. According to Darwin, how does isolation favor the creation of new varieties and species?

8. How does Darwin explain the fact that island populations tend to diminish or go extinct when the islands are invaded by species from the continents?

9. On July 1, 1858, two papers (see Link 4.3 below) were presented at a meeting of the Linnean Society in London, England—one written by Charles Darwin and the other by Alfred Russel Wallace—under the title "On the tendency of species to form varieties, and on the perpetuation of varieties and species by natural means of selection."

   Now that you've read the crux of Darwin's argument in Chapters 1–4 of this book, read Alfred Russel Wallace's paper on the same topic. How were his ideas similar to those of Darwin? How were they different?

Online Resources *available at* sites.sinauer.com/readabledarwin

Videos

4.1 Bird-of-Paradise Project from the Cornell Lab of Ornithology

4.2 Bumblebee gathering nectar from flowers

4.3 Platypus on land and in water

4.4 Lungfish breathing outside water and hibernating for years

4.5 A day in the life (and death) of a lungfish

4.6 Natural selection and adaptation in pocket mice

Links

4.1 The evolution of four-footed terrestrial animals from aquatic ancestors
www.evolution.berkeley.edu/evolibrary/article/evograms_04

4.2 More information (with photographs) about the spurs on rooster legs
www.ohioline.osu.edu/vme-fact/0014.html

4.3 Darwin and Wallace papers presented at the Linnean Society
www.calacademy.org/darwin/pdfs/Darwin_Wallace_1858.pdf

(Note: Web addresses may change. Go to sites.sinauer.com/readabledarwin for up-to-date links.)

Bibliography

Downing, A. J. 1845. *The Fruits and Fruit Trees of America*. New York.

Gray, A. 1856. *A Manual of the Botany of the Northern United States: Second Edition; including Virginia, Kentucky, and all east of the Mississippi; arranged according to the Natural System. (The Mosses and Liverworts by Wm. S. Sullivant.) With fourteen plates, illustrating the genera of the Cryptogamia*. New York.

Pierce, J. 1823. A memoir on the Catskill mountains with notices of their topography, scenery, mineralogy, zoology, and economic resources. *The American Journal of Science and Arts* 6(1): 86–97.

5

Laws of Variation

Darwin wrote The Origin of Species *before the principles of genetics were understood. Gregor Mendel published his pea plant paper in 1866, about seven years after* The Origin *was first published, but Darwin apparently never found that paper, and nobody ever sent him the information. At the time it was published most people seemed to miss the general applicability of Mendel's findings anyway, so even those who read the paper didn't understand that his findings also applied to species with "blended" inheritance rather than just to the discrete traits that Mendel had been working with. Indeed, Mendel seems never to have tried to contact Darwin, even after having read a translation of his book, so that even Mendel must have missed seeing the connection between his work and what was described in* The Origin. *Thus, much of what is contained in Chapter 5 is Darwin trying his best to make sense of the "laws of variation" without having any understanding of what those laws actually were; he had no idea of how variation was generated, controlled, or passed on to offspring. Writing this chapter must have been an extremely frustrating experience for him. Still, it is fascinating to see Darwin wrestling with the issue, and to see how he categorizes the sorts of variability that need to be explained. In the first part of the chapter he talks about possible causes of variability, and later he talks about patterns in the distribution of variability. But clearly he is always thinking about how natural selection deals with variation, and believes that although other factors—such as the prolonged disuse of parts, or gradual acclimatization to particular climatic regimes—may also be at work, natural selection has probably played the major role in shaping the patterns of variation that we see in the current populations of all organisms, both plants and animals.*

I have previously sometimes spoken as if the variations—so common and affecting so many different features in animals and plants under domestication, and to a lesser degree also in nature—were due to chance. This is not the case, of course, but it serves to acknowledge quite plainly our ignorance of how each particular variation originates. Variability seems somehow related to the conditions of life to which each species has been exposed over several successive generations. But it is very difficult to decide how much changed environmental conditions, such as those of climate and food availability, have acted in shaping the patterns we see today. Certainly, the innumerable complete coadaptations of structure that we see between various organisms throughout nature cannot be attributed simply to changes in climate and food supplies. But when a variation is of the slightest use to any organism, we cannot tell how much to attribute to the accumulative action of natural selection, and how much to the direct effects of environmental conditions. For example, furriers are well aware that their target animals have thicker and better fur the further north they live; but who can say how much of this difference results from selection for the warmest-clad individuals over many generations, and how much to the direct action of cold weather on the growth of fur? Innumerable instances are known to every naturalist of species not varying at all even when living under the most opposite climates, which leads me to think that the direct action of the surrounding conditions plays a far smaller role in producing variation than does a natural tendency to vary, caused by factors of which we are now quite ignorant.

In one sense, the conditions of life may be said not only to cause variability, either directly or indirectly, but also to include natural selection, for environmental conditions determine whether this or that variety shall survive and leave offspring for future generations. When people do the selecting in breeding domesticated animals or plants and agricultural crops, we can clearly see that the two elements of change are distinct; variability is in some manner excited, but it is our deliberate selection of who will survive and who will mate that causes variations to accumulate in particular directions. Similarly it is selection that leads to the survival of the fittest in nature.

In this chapter I will consider the possible roles of changing environmental conditions in producing variation as well as the use and disuse of parts, and will discuss variations that are themselves correlated with other traits. I will also review the levels of variability encountered at different levels of organization and talk again about how the steady accumulation of such variations in certain directions leads to the patterns that we currently see in so many species of both animals and plants.

Effects of Increased Use and Disuse of Parts as Controlled by Natural Selection

From the facts alluded to in Chapter 1, I think there can be no doubt that increased use of certain parts in our domestic animals has strengthened and enlarged them, and disuse diminished them, and that such modifications have been inherited.[1] For organisms living in nature we have no standard

---

[1] Although the evidence seemed compelling to him at the time, Darwin was wrong on this point. We now know that physical changes caused by use or disuse are not passed along to offspring.

(A)

(B)

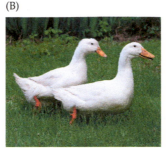

Figure 5.1 (A) Logger-headed duck (*Tachyeres brachypterus*) now better known as the steamer duck. These ducks are native to the Falkland Islands, slightly east of the southern tip of South America. (B) Aylesbury duck (rear) and drake (front).

of comparison by which to judge the effects of long-continued use or disuse, for we don't know what the organisms' ancestors looked like; but many animals possess structures that can best be explained by the effects of disuse. As Sir Richard Owen of the British Museum has remarked, there is no greater anomaly in nature than a bird that cannot fly; and yet there are several such bird species in this condition. The logger-headed duck of South America can only flap along the surface of the water, for example, and yet has its wings in nearly the same condition as the domestic Aylesbury duck (Figure 5.1); remarkably, according to the Scottish naturalist Mr. Robert Oliver Cunningham, while the adult birds cannot fly, the young birds can. As the larger ground-feeding birds seldom take flight except to escape danger, it seems likely that the nearly wingless condition of some birds that now inhabit or which recently inhabited several oceanic islands has been caused by disuse, since there are no predators on those islands. The ostrich, on the other hand, inhabits continents, not islands, and is certainly exposed to dangers from which escape by flight would be helpful; however, although it can defend itself quite well by kicking its enemies as efficiently as many four-legged animals, it cannot fly. It seems likely that the ancestor of the ostrich genus had habits like those of the large terrestrial birds known as bustards (Figure 5.2), and that, as the size and weight of its body increased during successive generations, its legs were used more often, and its wings less often until the birds became incapable of flight. I believe that natural selection has also played an important role here, and I will have more to say about this shortly.

William Kirby, coauthor of *An Introduction to Entomology*, has remarked (and indeed I have also observed this myself) that parts of the terminal section of the anterior legs (the "tarsi" or feet) of many

Figure 5.2 Kori bustard (*Ardeotis kori*) breeding display. Despite their size, bustard birds do fly.

Tarsus

Figure 5.3 Male dung beetle (*Onthophagus coenobita*).

male dung beetles (Figure 5.3) are often broken off; he examined 17 specimens in his own collection, and not one had even a relic of a foot remaining on the anterior legs. In the dung beetle species *Onites apelles*, the tarsi are so routinely lost that the insect has often been described as not having them. In some other genera they are present, but only in a rudimentary condition. In the sacred beetle of the Egyptians, a member of the genus *Ateuchus*, they are totally absent. How do we explain these facts? The evidence that accidental mutilations can be inherited is at present not decisive, but the remarkable cases observed by the French neurobiologist and physiologist Charles Édouard Brown-Séquard in guinea pigs, on the inherited effects of certain operations,[2] should make us cautious in denying this possibility. Perhaps it will be safest to look at the rudimentary condition of the anterior tarsi in some dung beetle genera and their entire absence in *Ateuchus*, not as cases of inherited mutilations but as due simply to the effects of disuse over a long period of time. Since many dung-feeding beetles are commonly found with their tarsi lost, this must happen early in life; obviously then, the tarsi cannot be of much importance to these insects.

In some cases we might incorrectly attribute modifications of structures to disuse when their reduction or absence is in fact wholly, or at least primarily, due to natural selection. The entomologist Mr. Thomas Vernon Wollaston has discovered the remarkable fact that of the more than 550 beetle species found on the island of Madeira, 200 species have such poorly developed wings that they cannot fly, and that of the 29 genera endemic to Madeira, 23 have all their species in this condition! How can we explain these remarkable facts? Well, we know the following: 1) that beetles in many parts of the world are frequently blown out to sea, where they perish; 2) that the beetles of Madeira, as observed by Mr. Wollaston, spend their time in hiding until the wind dies down and the sun shines; 3) that the proportion of wingless beetles is even larger on the fully exposed neighboring rock island of Desertas than on Madeira itself; and—a most extraordinary fact—4) that certain large groups of beetles that are very numerous elsewhere, and which absolutely require their wings to function, are almost entirely absent on Madeira. These facts convince me that the wingless condition of so many Madeiran beetles is mainly due to the action of natural selection, probably combined with the direct effects of disuse: during many successive generations, any individual beetles that flew the least, either from their wings having been ever so slightly less perfectly developed or from indolent

---

[2] In a series of experiments, Brown-Séquard found that severing the spinal cord or certain nerves in adult guinea pigs caused what seemed to be inherited problems for the offspring.

(A)   (B)

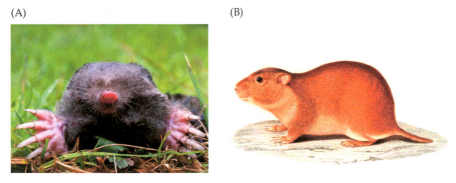

Figure 5.4 (A) European mole (*Talpa Europaea*); note how poorly developed the eyes are. (B) Brazilian tuco-tuco (*Ctenomys brasiliensis*); Darwin talks here about seeing this rodent when he was in Uruguay, during his voyage on the H.M.S. *Beagle*.

habit, will have had the best chance of surviving, simply from not having been blown out to sea. In contrast, those beetles that most readily took to flight would have been blown out to sea the most often and been destroyed.

The insects of Madeira that are not ground feeders and which, as with certain flower-feeding coleopterans (beetles) and lepidopterans (moths and butterflies), must use their wings to find food have, as Mr. Wollaston suspects, fully developed—even enlarged—wings. This is quite compatible with the action of natural selection. For when a new insect first arrived on the island, the tendency of natural selection to enlarge or to reduce the wings would depend on whether more individuals were saved by successfully battling with the winds or by giving up the attempt and rarely, if ever, flying. Similarly, as with sailors shipwrecked near a coast, it would have been better for the good swimmers if they had been able to swim further, whereas it would have been better for the bad swimmers if they had not been able to swim at all, and had instead stayed with the wreck.

A similar series of gradations can be seen among rodents and moles. The eyes of moles (Figure 5.4A) and of some burrowing rodents are rudimentary in size, and in some cases are covered by skin and fur. This state of the eyes is probably due to gradual reduction from disuse, but aided perhaps by natural selection. In South America, the tuco-tuco (Figure 5.4B), a burrowing rodent in the genus *Ctenomys* that is even more subterranean in its habits than the mole, is frequently blind. One that I kept alive myself was certainly blind; on dissection I found the cause to have been inflammation of the nictitating membrane, which is basically a transparent or translucent, supplementary eyelid. As frequent inflammation of the eyes must be injurious to any animal, and as eyes are certainly not needed by animals living below the ground, a reduction in their size, with the adhesion of the eyelids and growth of fur over them, might in such cases be an advantage; if so, then natural selection would augment the effects of disuse.

It is well known that several animals inhabiting the caves of both Carniola[3] in Europe and the American state of Kentucky in North America—and belonging to a wide range of different taxonomic classes—are blind. Some of the blind crabs in these caves retain their eye stalks, although the eye itself is gone: the *stand* for the telescope remains, but the telescope itself, with its lenses, has been lost. As it is difficult to imagine that eyes, even though useless, could in any way be injurious to animals living in darkness, their loss may be attributed to simple disuse. Professor Benjamin Silliman of Yale University captured two blind cave rats (genus *Neotoma*) more than a half mile from the mouth of a cave, thus in darkness but not in the profoundest depths. Their eyes were lustrous and large; after exposing the animals to light of gradually increasing intensity for about one month, Professor Silliman tells me that the animals acquired a dim perception of objects. Clearly the eyes remained at least partly functional.

It is hard to imagine conditions of life more similar than that in deep limestone caverns under nearly similar temperature and humidity. In accordance with the traditional view that each of the various blind animals has been separately created for the American and European caverns, we would thus expect to find many similarities in their organization and affinities. But if we look at the two whole faunas, we see that they are *not* similar in organization or affinity. With respect to insects alone, the Danish entomologist Jørgen Matthias Christian Schiødte has remarked that they seem "purely local," and that "…the similarity that is exhibited in a few forms between the Mammoth Cave in Kentucky and the caves in Carniola" is simply "a very plain expression of that analogy which subsists generally between the fauna of Europe and of North America." In my view, we must suppose that American animals with ordinary powers of vision slowly migrated over successive generations from the outer world into the deeper and deeper recesses of the Kentucky caves, and that European animals independently migrated into the caves of Europe. We do, in fact, have some evidence of this gradation of habit. As Mr. Schiødte remarks, "We accordingly look upon the subterranean faunas as small ramifications which have penetrated into the earth from the geographically limited faunas of the adjacent tracts, and which, as they extended themselves into darkness, have been accommodated to surrounding circumstances. Animals not far remote from ordinary forms prepare the transition from light to darkness. Next follow those that are constructed for twilight; and last of all, those destined for total darkness, and whose formation is quite peculiar." It is important to realize that Mr. Schiødte's remarks apply not to the individuals of any one species, but rather to different species. By the time an animal has reached, after numberless generations, the deepest recesses, disuse will, on this view, have more or less perfectly obliterated its eyes, and natural selection will often then

---

[3] The area known then as Carniola is essentially what we know today as Slovenia.

have brought about other changes, such as an increase in the length of the antennae or sensory palps,[4] as compensation for blindness.

Notwithstanding such modifications, if they have indeed been brought about by natural selection, we might still expect to see some affinities between the cave animals of North America and other non-cave-inhabiting animals of that continent, and, likewise, between the cave animals of Europe and other inhabitants of the European continent. This is indeed the case with at least some of the American cave animals, as I hear from Dr. James Dwight Dana at Yale University. Similarly, some of the European cave insects are very closely allied to the non-cave insects of the surrounding areas in Europe. It would be difficult to explain the affinities of the blind cave animals to the other inhabitants of the two continents on the traditional view of the cave animals having been specially and independently created. We might instead have expected to see a close relationship between those inhabiting the caves of the Old and New Worlds, from the well-known close relationships among most of their other fauna. A blind genus of ground beetles (*Anophthalmus*) offers us a particularly good, although rare, example of such close relationships: as the entomologist Mr. Andrew Murray observes, although the species has so far been found only in caves, those that inhabit the several caves of Europe and America resemble each other. Perhaps the ancestors of these several species, whilst they were still furnished with eyes, had formerly ranged over both continents and then become extinct, except in their present secluded abodes. Far from being surprised that some cave animals should lack any closely-related terrestrial counterparts, as Louis Agassiz of Harvard University has found with regard to the blind American cave fish in the genus *Ambylopsis*, and as is also the case with the blind, cave-dwelling, European amphibian *Proteus*, I am only surprised that more such wrecks of ancient life have not been preserved in these caves, owing to the less severe competition to which the very few inhabitants of these dark abodes will have been exposed.

Acclimatization

The period of flowering in plants, the season in which the plants become dormant, the amount of rain needed for seeds to germinate—these and similar habits are all hereditary with plants. This leads me to say a few words about acclimatization. As it is extremely common for distinct species belonging to the same genus to inhabit both hot and cold countries, then if it be true that all members of the same genus are descended from a single ancestor, acclimatization to different climates must happen easily over many generations.

---

[4] These are sensory appendages found on the heads of crustaceans and insects, formed from certain of the mouthparts.

It is notorious that each species is adapted to the climate of its own home: species from an arctic or even from a temperate region cannot endure a tropical climate, and conversely. Similarly, many succulent desert plants cannot endure a damp climate. But the degree to which species are adapted to the climates under which they live is often overrated.

We may infer this from our frequent inability to predict whether or not any particular imported plant will endure our climate here in England, and from the number of plants and animals brought from different countries that do perfectly well here. We have reason to believe that species in nature are tightly restricted in their ranges at least as much by competition with other organisms as by their being adapted to particular climates. Indeed, we have clear evidence that some plants have sometimes become, to at least a certain extent, well habituated to new and different conditions. Thus the excellent botanist (and my good friend) Dr. Joseph Hooker has collected seeds from certain species of pines and rhododendrons (Figure 5.5) growing at different heights in the Himalayas of Asia and found that when the seeds were planted in this country, plants from the same species collected at different heights on the mountain differed in their ability to resist cold. Another botanist, Mr. George Henry Kendrick Thwaites, has observed similar facts in Ceylon, and Mr. Hewett Watson has made analogous observations about European species of plants brought from the Azores to England. There are more examples that I could give. In regard to animals, several authentic instances could be adduced of species having largely extended their range from warmer to cooler latitudes within historical times, and conversely. We do not know with certainty that these animals were strictly adapted to their native climate, though in all ordinary cases we assume that to be the case. Nor do we know whether they have subsequently become specially acclimatized to their new homes, so as to be better fitted for them than they were at first.

It seems reasonable to assume that our domestic animals were originally chosen by uncivilized man because they were useful in some way and because they reproduced readily under confinement, and not because they were subsequently found capable of being transported large distances. The common and extraordinary capacity of our domestic animals to not only be able to withstand a wide range of climates

Figure 5.5 Himalayan rhododendrons.

but of also being perfectly fertile (a far more severe test) under such a wide range of conditions thus suggests that a large proportion of other animals now living in nature could also easily be brought to withstand widely different climates. We must not, however, push the foregoing argument too far, remembering that some of our domestic animals originated from several wild stocks. The blood, for instance, of a tropical and arctic wolf may perhaps be mingled in our domestic breeds. Although the rat and mouse cannot be considered as domestic animals, we certainly have transported them to many parts of the world so that they now have a far wider geographical range than any other rodent; they live under the cold climate of the Faroe Islands in the North Atlantic and of the Falkland Islands in the South Atlantic, and yet also on many an island in the intensely hot tropics. Thus adaptation to any special climate may be looked at as a quality readily grafted onto an innate flexibility of constitution that is common to most animals. Viewed thus, the ability to endure the most different climates by man himself and by his domestic animals, and the fact of the extinct elephant and rhinoceros having previously endured a glacial climate, whereas the living species are now all tropical or sub-tropical in their habits, ought not to be looked at as anomalies, but rather as examples of a very common flexibility of constitution, brought into action under peculiar circumstances.

How much of the acclimatization of species to any peculiar climate is due to mere physiological flexibility of individuals and how much to natural selection acting on varieties with different innate constitutions over time— or to both of these mechanisms combined—is an obscure question. It seems reasonable that both physiological flexibility and experience have had some influence. And as it is not likely that humans should have succeeded in selecting so many breeds and subbreeds with constitutions specially fitted for their own regions, the result must, I think, be due to physiological flexibility. On the other hand, natural selection would inevitably tend to preserve those individuals that were born with constitutions best adapted to whatever country they inhabited. In treatises on many kinds of cultivated plants, certain varieties are said to withstand certain climates better than others. This is strikingly shown in works published in the United States about fruit trees: certain varieties are habitually recommended for the northern states and others for the southern states; as most of these varieties are of recent origin, they cannot owe their constitutional differences to habit. The case of the Jerusalem artichoke, which is never propagated in England by seed, and of which consequently new varieties have never been produced, has even been advanced as proving that acclimatization cannot be achieved, for it is now as tender as it ever was! The case of the kidney bean has also been cited for a similar purpose, and with much greater weight; but until someone will plant his kidney beans so early that a very large proportion of them are destroyed by frost and then collect seed from the few survivors, with care to prevent accidental crosses, and then again get seed from those seedlings with the same precautions over a dozen or more generations, then the experiment cannot be said to have been tried.

Nor let it be supposed that differences in the constitution of seedling kidney beans never appear, for I have read that some seedlings are hardier than others, and in fact I have observed striking instances of this in my own studies. It may well be possible to select for cold tolerance in kidney beans.

On the whole, we may conclude that physiological flexibility and experience along with use and disuse, have, in some cases, all played a considerable part in the modification of the constitution and structure of animals and plants. **But clearly these effects have often been largely combined with—and sometimes overmastered by—natural selection acting on innate variation among individuals within species.**

Correlated Variation

By "correlated variation" I mean that all of the parts of an organism are so tied together during its growth and development that when slight variations in any one part occur, and are gradually accumulated by natural selection, some other, apparently unrelated parts become modified as well. This is a very important subject, but one that we understand very poorly. We shall presently see that what seems like correlation in many other cases can be easily explained by simple inheritance.

One of the most obvious legitimate cases of correlated variation is when anatomical variations arising in the young or larvae of a species later affect the structure of the mature animal. The several parts of the body that are symmetrical, and which at an early embryonic period are identical in structure and are necessarily exposed to similar environmental conditions, seem eminently liable to vary in similar ways: we see this in the right and left sides of the body varying in the same manner, for example. The front and hind legs of many species also vary together. However, these tendencies may be overruled in some cases more or less completely by natural selection. For instance, a family of stags once existed with an antler only one side of the head. If this situation had been of any great use to the breed it would probably have been rendered permanent by selection, and the left and right sides in that species would no longer be "correlated."

It also seems that hard body parts can affect the form of adjoining soft body parts, and that may account for the apparent correlation of some traits. Some authors believe that the diversity in the shape of the pelvis among birds causes the remarkable diversity in the shape of their kidneys. Others believe that in humans, the shape of the child's head is influenced, through pressure, by the shape of the mother's pelvis. And in snakes, according to the German naturalist Hermann Schlegel, the form of the body and the manner of swallowing determine the position and form of several of the most important digestive organs.

However, the nature of the relationship is frequently quite obscure. The French authority on developmental abnormalities Isidore Geoffroy Saint-Hilaire[5] is firmly convinced that certain physical malformations tend to appear

---

[5] Isidore Geoffroy Saint-Hilaire is the son of French naturalist Étienne Geoffroy Saint-Hilaire.

Peripheral flower

Open central flower

Unopened central flower

Figure 5.6 Daisy (*Leucanthemum vulgare*) showing peripheral and central flowers.

together, although what causes this connection remains uncertain. What can be more singular than the relationship in cats between completely white fur, blue eyes, and deafness, or between the tortoise-shell color in cats and being female; or with pigeons that have feathered feet also having skin between the toes, or with the amount of down on the skin of young pigeons and the future color of its plumage? And I think it can be hardly accidental that the two orders of mammals that are most unusual in their skin coverings, that is, the Cetacea (whales, dolphins, and porpoises) and the Edentata[6] (armadillos, anteaters, and sloths, for example) are also the most abnormal in their teeth.

Perhaps the best example showing the importance of the laws of correlation and variation—unrelated to usefulness and therefore not affected by natural selection—is that of the difference between the outer and inner flowers in some members of the Asteracea (or Compositae) family of plants (including daisies, asters, and sunflowers), and those in the Apiaceae family (including carrots and parsley). These plants all produce clusters of small flowers that give the appearance of being a single flower. For example, everyone is familiar with the peripheral ray flowers and the central florets of the daisy, where the small clustered flowers form a central disk (Figure 5.6); that is, although the daisy looks like a single flower, it is really composed of many individual flowers, with those on the outside (the peripheral flowers) looking very different—each peripheral flower produces a single long petal—from those more centrally located. This difference is often associated with the ray flowers being either sterile (having neither male nor female parts) or forming only female parts. But in some of these plants, the seeds of the different flowers also differ in shape and sculpture. These differences have sometimes been attributed to the pressure placed on the florets[7] by the involucra—the group of small green leaflets at the base of each flower cluster; the shape of the seeds in the peripheral florets of some members of the Asteracea (daisies, for

---

[6] These animals are now all placed within the order Xenarthra.

[7] One of the small flowers that makes up the head of a composite flower.

example) support this idea. On the other hand, with plants in the Apiaceae family (carrots and parsley), Dr. Hooker tells me that the species with the densest heads do *not* differ most frequently in their inner and outer flowers. It might have been thought that by drawing their nourishment from the reproductive organs, the ray petals cause the abortion of those organs; but this cannot be the sole cause, for in some members of the Asteracea the seeds of the outer and inner florets differ even though there is no difference in the petals. Possibly these several differences may be connected with the different flow of nutrients toward the central versus the peripheral flowers; we know that with irregular flowers, those nearest to the axis are most likely to become abnormally symmetrical. I may add, as an example of this fact, and as a striking correlation, that in many geraniums, the two upper petals in the central flower of the cluster often lose their patches of darker color; and when this occurs, the adhering nectary—which of course produces the sugar-rich nectar—is quite aborted. The central flower thus becomes unusually symmetrical. When the color is absent from only one of the two upper petals, the nectary is much shortened, although not fully aborted.

With respect to the development of the corolla of these composite flowers, Christian Konrad Sprengel's idea that the peripheral florets serve to attract insects is highly likely, as insects are highly advantageous, or even essential, for fertilizing these plants; if so, then natural selection has probably been at work. But with respect to the seeds, it seems impossible that their shape differences, which are not always correlated with any difference in the corolla, can in any way benefit the plant; yet, in the carrot/parsley family (the Apiaceae), the seeds are sometimes very straight in the exterior flowers but hollow and curved in the central flowers. Indeed, the Swiss botanist Augustin de Candolle (the father of Alphonse de Candolle, see below), based his main taxonomic divisions for the entire order on such characteristics. Thus we see that modifications of structure, viewed by systematists[8] as of great importance, may be wholly due to the laws of variation and correlation, without being, as far as we can judge, of the slightest benefit to the species or subject to the laws of natural selection.

Although correlated variation is common, certain structures that are common to whole groups of species exist not through correlation but rather to inheritance. An ancient ancestor may have acquired, through natural selection, some particular modification in structure and, after thousands more generations, some other and independent modification. These two modifications, having been transmitted to a whole group of descendants with diverse habits, would naturally be thought to be in some necessary matter correlated, even though they were actually independently evolved.

---

[8] Systematists are biologists who try to understand how different groups of organisms are related to each other.

Some other correlations are apparently due to the manner in which natural selection alone can act. For instance, the younger de Candolle (Alphonse) has noted that winged seeds are never found in fruits that do not open. I should explain this correlation as follows: it would be impossible for seeds to gradually become winged through natural selection unless the capsules surrounding them eventually opened to the outside, releasing the seeds; only in that way could seeds that were a little better adapted to be wafted by the wind gain an advantage over others less well fitted for wide dispersal. In short, through natural selection, we would never expect winged seeds to evolve in plants that never released their seeds to the wind.

Compensation and the Economy of Growth

Both Étienne Geoffroy Saint-Hilaire (Isidore Geoffroy Saint-Hilaire's father) and Johann Wolfgang von Goethe formulated, at about the same time, their "law of compensation." As Goethe expressed it, "…in order to spend on one side, nature is forced to economise on the other side." I think this also holds true to a certain extent with our domesticated species: if nourishment flows to one part or organ in excess, it rarely flows to another part, at least to such a high degree. For example, it is difficult to get a cow to give a lot of milk and to fatten it readily at the same time. Similarly, some varieties of cabbage yield an abundant and nutritious foliage but only a few seeds, while other varieties produce a copious supply of oil-bearing seeds, but a less-developed foliage. And again, when the seeds in our fruits become atrophied, the fruit itself increases considerably in size and quality, again suggesting an increase of nutrients in one part being caused by a decrease in another part. Similarly, a large tuft of feathers on the heads of our poultry is generally accompanied by a smaller comb on the top of the head, and a large beard is similarly accompanied by diminished wattles under the chin.

In nature it is difficult to tell whether this "law of compensation" applies or not. I don't see any way of distinguishing between the effects of a part being largely developed through natural selection—with another and adjoining part being reduced by the same process—on the one hand, and the excess growth of some part directly causing a withdrawal of nutrients from some other part, on the other hand.

I also suspect that some of the presumed cases of compensation that have been proposed in the literature may be combined with some other facts under the more general principle that natural selection is continually trying to economize every part of an organism's organization. If under changed environmental conditions a structure that was previously useful now becomes less useful, its diminution and eventual loss will be favored, for it will benefit the individual not to have its nutrients wasted in trying to build and maintain a useless structure. This is the only way that I can understand a fact with which I was much struck when examining barnacles some years ago, namely that when one barnacle is parasitic within another, and is thus protected, it more or less completely loses its own outer covering, or shell. This is certainly the case with the males of the barnacle genus *Ibla*, and is seen in a truly extraordinary manner within the genus *Proteolepas*. The outer covering in the larvae of all other barnacles consists of three highly important anterior segments of the head enormously developed and furnished with large nerve fibers and muscles; but in members of the parasitic and protected genus *Proteolepas*, the whole anterior part of the head is reduced to the merest rudiment[9] attached to the bases of the prehensile antennae.

Many analogous cases could be given. Clearly it would be a decided advantage to each individual of any species not to produce a large and complex structure that has now become superfluous; in the struggle for life that all animals experience, individuals will have a better chance of supporting themselves by wasting less of what they eat. Thus I believe that natural selection will tend in the long run to reduce any part of the organization that becomes, through changed habits, superfluous, without causing some other part to be largely developed in a corresponding degree. Conversely, natural selection may also perfectly well succeed in increasing the size of an organ if that should prove beneficial, without requiring the compensatory reduction of an adjoining part.

Multiple, Rudimentary,[10] and Lowly Organized Structures Are Especially Variable

As the French zoologist Isidore Geoffroy Saint-Hilaire (the son of Étienne, as mentioned earlier in the chapter) has noted, when any part or organ occurs many times in the same individual (e.g., the vertebrae of snakes, or the stamens in polyandrous[11] flowers (i.e., those having numerous stamens), their number usually varies among individuals; in contrast, when the same part or organ occurs within an organism in fewer numbers, the number is constant from one individual to the next. Mr. Étienne Geoffroy Saint-Hilaire, together with a number of botanists, has noted that multiple parts in flowers are also extremely likely to vary in structure. This observation fits well with Sir Owen's suggestion that the simple repetition of parts reflects what he calls "low organization"; as many naturalists have suggested, more primitive organisms tend to vary more than do more advanced organisms. This makes sense to me if by "primitive" we mean that the various parts of an organism's organization have not become specialized for particular functions: as long as the same part has to perform many different tasks, we can

---

[9] A rudiment is the barest beginning of a structure. Think of "rudimentary knowledge." In this case the head only starts to develop, but never gets very far.

[10] We now refer to such structures as "vestigial," a mere remnant of something formerly better developed.

[11] The Greek derivation of polyandrous: poly = many; and androus = husbands!

perhaps see why it should remain variable—that is, why natural selection would not have preserved or rejected each little deviation of form as carefully as when the part has had to serve some specialized purpose. In the same way, a knife that has to cut all sorts of very different things may be almost any shape, whereas a cutting tool used for some specific purpose will be of some particular shape. **It should never be forgotten that natural selection can act only through and for the advantage of each being.**

It is also widely acknowledged that rudimentary parts of all sorts tend to be highly variable in form. I will return to this topic later. Here let me just say that their great variability among individuals seems to result from their uselessness, and thus from natural selection having no power to stop deviations in their structure from persisting.

A Part Developed in Any Species to an Extraordinary Degree Tends to Be Highly Variable, in Comparison with the Same Part in Related (Allied) Species

Several years ago I was much struck by the English naturalist Mr. George Robert Waterhouse's remark that parts developed to an extraordinary degree in some species tend to be far more variable than the same parts developed normally in related species. Sir Owen seems to have come to a similar conclusion in his studies. I am convinced that this is in fact a rule of great generality, although I cannot possibly introduce here the long array of facts that I have collected that have led me to this conviction. It should be understood that the rule applies only to parts that are unusually well developed in one species or in a few species in comparison with the same part found in many closely allied species. Thus, although a bat's wing is a most abnormal structure to find among mammals, the rule would not apply here because all bats have wings; it would apply only if one particular bat species had wings that were remarkably different in some way in comparison with those of the other species in the same genus.

The rule applies very strongly in the case of secondary sexual characteristics, a term which, as used by the Scottish anatomist, John Hunter, refers to characteristics associated with only one sex without being directly connected to the physical act of reproduction—such as the long and fantastically colored feathers of the male peacock (Figure 5.7). The rule

Figure 5.7 A beautiful Indian peacock (*Pavo* sp.) with fully fanned tail.

(A)

(B)

(C)

Figure 5.8 (A) Underwater photo of acorn barnacle (*Balanus glandula*). (B) The pontellid copepod (*Pontella securifer*). Various parts glow fluorescent green when viewed under blue light. Top: Note the male's hinged antenna and the claw on its last leg, both used for grabbing females. (C) Freshwater copepod (*Cyclops* sp.).

applies to both males and females, although males are more likely to show such secondary sexual characteristics, and in a highly variable manner. But our rule is not confined to such characteristics, as is clearly shown in the case of hermaphroditic barnacles; indeed, the rule almost always holds good for these interesting animals, which I have studied in detail for many years. In a future work I will list all the more remarkable cases. Here I will give only one example, but one that illustrates the rule in its largest application.

Rock barnacles have very important outer body parts called "opercular valves"—which are calcareous structures (i.e., containing calcium carbonate) that are moved together or apart to open and close a barnacle's shell (Figure 5.8A). Typically they differ extremely little even among the members of distinct genera. However, in several species of the barnacle genus *Pyrgoma*, these valves present a marvelous amount of diversification. The comparable valves found among the different species of that genus are sometimes completely different in shape, and the amount of variation among the individuals of even the same species is so great that I can say without exaggeration that *varieties* of the same species differ more from each other in the characteristics derived from these important organs than they do among species belonging to other genera.

Over the years I have also paid particular attention to birds, as individuals of the same species inhabiting a common area generally vary extremely little; I find that the rule seems to hold up very well in this class of animals. I can't determine how well it applies in plants, though; the great variability in plants makes it particularly difficult to compare their relative degrees of variability.

Whenever we see any part or organ of a species developed to a very remarkable degree in any particular species, it is fair to presume that that part has a highly important function. And yet, it is eminently liable to variation among individuals. Why should this be? On the view that each species has been independently created with all of its parts just as we see them now, I can see no explanation. But if we assume that groups of species are descended from other, ancestral species, and have been modified over time by natural selection, then I think the observation makes good sense.

First let me make some preliminary remarks. If any part in one of our domesticated animals—or even the whole animal—be neglected during selection for breeding future generations, with no selection being applied in any direction, then that part—or the entire breed—will soon cease to have a uniform character: it may be said that the breed is degenerating. We see a nearly parallel case in rudimentary organs and in those that have not been specialized for any particular purpose: in such cases, natural selection has not—or cannot—come into full play, so that the organization is left in a fluctuating condition.

But what especially concerns us here is that the features in our domestic animals that are presently undergoing rapid change due to deliberate selection of traits in breeding are also eminently liable to variation within each breed. Look, for example, at individuals of the same breed of pigeon and see what a prodigious amount of variation there is in the beak of tumblers (see Figure 1.2), in both the beak and the wattle of carriers (see Figure 1.3A), in the carriage and tail of fantails (see Figure 1.3C), and so on, which are all currently features of major concern to English pigeon fanciers (i.e., breeders). Even in the same subbreed of pigeon, as in that of the short-faced tumbler, it is notoriously difficult to breed nearly perfect birds; many depart widely from the standard. Truly, there is a constant struggle between the tendency to revert to a less perfect state along with an innate tendency to new variations on the one hand, and the power of steady selection to keep the breed true on the other. In the long run, selection wins the day: we do not expect to fail so completely as to breed a bird as coarse as a common tumbler pigeon from a good short-faced strain. But as long as active selection is going on, we can always expect to see much variability in the parts undergoing modification.

Now let us turn to nature. Whenever any particular part has been developed to an extraordinary manner in any one species, compared with the development of that part in other species in the same genus, we may conclude that this part has undergone an extraordinary amount of modification since the period when the several species within that genus branched off from their common ancestor. This period will seldom be extremely remote, as species rarely endure for more than one geological period. An extraordinary amount of modification implies an unusually large amount of variability continued over a long period of time that has been continually accumulated for the benefit of the species by natural selection. But as the variability of the extraordinarily developed part or organ has been so great in the fairly recent past, we might, as a general rule, still expect to find more variability in such parts than in other parts of the organisms that have remained nearly constant for a much longer time. This is, I am convinced, precisely the case. I see no reason to doubt that the struggle between natural selection and the tendency to reversion and variability will eventually cease, and that the most abnormally developed organs may then be made constant. Thus, when an organ, however abnormal it may seem to us, has been transmitted in approximately the same condition to many modified descendants over many generations—as in the case of the bat's wing—it must, according to our theory, have existed for an immense period of time in nearly the same state; thus it is now no more variable than any other structure. It is only in those cases in which the modification has been extraordinarily great and comparatively recent that we would expect to find considerable variability still present.

Specific Characters Vary More Than Generic Characters

The same principle discussed in the preceding section may also explain the differences in variability in specific and generic characteristics. It is notorious that specific traits vary more than generic ones do. For example, suppose that in a large genus of plants, some species had blue flowers while some had red flowers; flower color would then be only a species-defining characteristic and no one would be surprised at seeing one of the blue species varying into red, or one of the red species varying into blue within the genus. But if all the species in that genus had blue flowers, the color would become a generic character and its variation would be a more unusual circumstance. I have chosen this example because most naturalists would argue that specific characteristics are more variable than generic ones simply because they are taken from parts of less physiological importance than those commonly used for classing genera. I believe this explanation is only partly, and indirectly, true; I shall return to this point in the chapter on classification[12] (Chapter 14). With respect to particularly *important* characteristics I have repeatedly noticed in works on natural history that when some important organ or part that is generally very constant throughout a large group of species differs considerably in some closely allied species, it is often variable among individuals of some of the species. This shows that when a trait that is usually used to define genera sinks in value and becomes only useful in defining species, it often becomes variable, even though its physiological importance to the organism may remain the same. **In other words, characteristics that vary among species also vary greatly among individuals of those species, regardless of their physiological importance.**

---

[12] Darwin's cross-references to chapters beyond Chapter 8 have been retained in this volume so that readers can refer to the original *The Origin of Species*, Sixth Edition.

On the ordinary view of each species having been specially and independently created, why should one part of a structure that differs from the same part in other independently created species of the same genus be more variable than those parts that are very similar in the same species? I do not see that any explanation can be given. **But if we believe that species are only strongly marked and fixed varieties, then we might expect to often find them still continuing to vary in those parts of the structure that have varied in the recent past.** To state the case in a different way, the characteristics in which all the species of a genus resemble each other, and in which they differ from allied genera, are called generic characteristics and are used to define the members of a genus; those characteristics may be attributed to inheritance from a common ancestor, for it can rarely have happened that natural selection will have modified several unrelated species, fitted to more or less widely different habits, in exactly the same way. And as these so-called generic characteristics have been inherited from before the time when the various species first branched off from their common ancestor, and subsequently have not come to differ to any large extent, it is not likely that they should vary today. On the other hand, since the characteristics that distinguish one species from another within the same genus (specific characteristics) have varied and come to differ *after* the time when the various species branched off from their common ancestor, it is reasonable that they should still often be somewhat variable today—at least more variable than those parts of the organism that have remained constant for a very long period.

Secondary Sexual Characteristics Are Highly Variable

I think most naturalists will admit that secondary sexual characteristics are highly variable. It will also be admitted that species belonging to any particular group differ from each other more widely in their secondary sexual characteristics than in other parts of their anatomy. Consider, for example, the turkey, chicken, pheasant, and other gallinaceous[13] birds, and compare, for instance, the amount of difference between the males, in which secondary sexual characteristics are strongly displayed, with the far more limited differences between the females. Although we don't know what causes the original variability of those characteristics, we can easily see why they should not have been made as constant and uniform as other traits: they are accumulated by sexual selection, which is less rigid in its action than ordinary selection since it does not involve death, but only gives fewer offspring to the less-favored males. As secondary sexual characteristics are highly variable, sexual selection will have had a wide scope for action, and may thus have succeeded in giving to the species of the same group a greater amount of difference in these traits than in others.

Remarkably, the differences between the two sexes of the same species are generally displayed in the very same parts of the anatomy in which

---

[13] Gallinaceous refers to birds of the order Galliformes.

species of the same genus differ from each other. Let me just give the first two examples on my long list of examples to illustrate this point. As the differences in these cases are of a very unusual nature, the relation can hardly be accidental.

Among most groups of beetles, the terminal section of the leg is called the tarsus, or foot. For beetles in most groups, all individuals have the same number of segments in the tarsi. But in the beetle family Engidae,[14] as the entomologist John Obadiah Westwood has remarked, the number varies greatly, and even more intriguing, the numbers differ in the two sexes of each species. Again, in the ground-dwelling Hymenoptera (a very large group of species that includes the wasps, bees, and ants), the pattern of nerves in the wings is a taxonomic characteristic of the highest importance, because it is common to large groups; but in certain genera the pattern differs among different species, as it also does between males and females of the same species. The very versatile archaeologist and biologist Sir John Avebury Lubbock has recently remarked that several minute marine crustaceans—in the copepod genus *Pontella*—offer excellent illustrations of this law (see Figure 5.8B, C). "In *Pontella*, for instance, the sexual characters are afforded mainly by the anterior antennae and by the fifth pair of legs: the differences between species are also principally given by these organs."[15] This relation has a clear meaning in my view of things: I look at all the species of the same genus as having certainly descended from a common ancestor, just as have the two sexes of any one species. Consequently, whatever part of the structure of the common ancestor, or of its early descendants, became variable, variations of this part would, mostly likely, be taken advantage of by both normal selection and sexual selection. While natural selection would act to gradually fit the different species to their particular niches, sexual selection would gradually act to fit the two sexes of the same species to each other, or to enable the males to best struggle successfully with other males for possession of the females.

Finally then, here are the key facts before us: 1) characteristics that distinguish species from each other are more variable than are those that distinguish the different genera from each other, or those that are possessed by all the species; 2) characteristics that are developed in a species in an extraordinary manner are frequently far more variable than the same parts in other species in the same genus; 3) parts that are common to a whole group of species show only a slight degree of variability, however extraordinarily those parts may be developed; 4) secondary sexual characteristics are highly variable, and show great differences in closely allied species; and 5) secondary sexual characteristics and ordinary specific differences are generally displayed in the same parts of the anatomy. These principles are all closely

---

[14] These beetles are now in the family Erotylidae.

[15] In male copepods, the right first antenna is always hinged, to grasp the female in mating, and the right limb of the fifth pair of hind legs is always shaped like a claw, for the same purpose. In some species, both the left and right limbs are so modified.

connected, all being mainly due to 1) all of the species of a particular group being descended from a common ancestor, from whom they have inherited much in common; 2) parts that have recently and largely varied being more likely to go on varying than parts that have long been inherited and have not varied; 3) natural selection having more or less completely, according to the lapse of time, overmastered the tendency to reversion and to further variability; 4) sexual selection being less rigid than ordinary selection; and 5) variations in the same body parts having been accumulated by both natural selection and sexual selection, and having been thus adapted for secondary sexual purposes as well as for ordinary, non-sexual purposes.

Distinct Species Present Analogous Variations, So That a Variety of One Species Often Presents a Trait Typical of a Related Species, or Reverts Back to Some Trait Possessed by an Early Ancestor

These claims will be most readily understood by considering our domestic animals and plants. In widely separated regions, the most distinct breeds of pigeon each sometimes present subvarieties with reversed feathers on the head or with feathers on the feet—characteristics never found in the aboriginal rock pigeon that gave rise to all pigeon breeds long ago. These then are "analogous variations" in two or more distinct races. The frequent appearance of 14 or even 16 tail feathers in some pouter pigeons (see Figure 1.2) may be considered as a variation representing the normal structure of another race of pigeons, the fantail. I presume that no one will doubt that all such analogous variations are due to the several races of the pigeon having inherited from a common parent the same constitution and the same tendency to vary in certain directions, when acted on by similar (but presently unknown) influences.

Among plants we have a similar case of analogous variation in the enlarged stems of the Swedish turnip and the rutabaga (Figure 5.9), plants that several botanists rank as mere varieties produced by cultivation from a common parent. If this is not the case, then it represents analogous variation in two so-called distinct species; and to these then, a third species may be added, namely the common turnip (Figure 5.10), which shows the same enlarged stems. If we are to believe that each species was independently created, we should have to attribute this similarity in the enlarged stems of these three plants not to descent from a common ancestor and a consequent tendency to vary in a similar

Figure 5.9 Rutabaga (*Brassica napus*).

Figure 5.10 Common turnip (*Brassica rapa*).

manner, but rather to three separate yet closely related acts of creation. Many similar cases of analogous variation have been observed by the French botanist Charles Victor Naudin in the great gourd family, and by various other authors in wheat and other cereals. Similar cases occurring with insects under natural conditions have recently been discussed by the American entomologist Mr. Benjamin Walsh.

With pigeons, we have a particularly striking case, namely the occasional appearance of slaty-blue birds with two black bars on the wings, white loins, a bar at the end of the tail, and with the outer feathers externally edged with white near their base; such characteristics occasionally appear in all pigeon breeds. As these marks all characterize the parent rock pigeon (see Figure 1.3D), I presume no one will doubt that this is a clear case of reversion to ancestral characteristics, and not of a new analogous variation suddenly appearing in the several breeds. We may, I think, confidently come to this conclusion because, as we have seen earlier, these colored marks are eminently liable to appear in the offspring when two distinct and differently colored breeds are mated; there is nothing in the external conditions of life to cause that reappearance of the slate-blue coloration, or with the several other marks described, beyond the influence of the mere act of crossing on the laws of inheritance, whatever they may turn out to be.

No doubt it is a very surprising fact that characteristics should sometimes reappear after having been lost for many—indeed, probably for many hundreds of—generations. But when a breed has been crossed only once with some other breed, the offspring occasionally show a tendency to revert in character to the foreign breed for many generations—some say for a dozen or even a score of generations. This is quite remarkable; after 12 generations, the proportion of blood, to use a common expression, from one ancestor is only 1 part in 2,048; and yet, as we see, it is generally believed that this remnant of foreign blood somehow retains a tendency to reversion.[16] In a breed that has not been crossed, but in which both parents have lost some particular characteristic that their ancestor possessed, the tendency, whether strong or weak, to reproduce that lost character might, as was formerly remarked, be transmitted to offspring for almost any number of generations. When a characteristic that has been seemingly lost in a breed

---

[16] We now know that the traits Darwin is talking about are controlled by particular recessive alleles—the forms of a gene that are expressed only when paired with an identical allele, and are not expressed when paired with a dominant allele—not by anything in the blood.

reappears after a great number of generations, the most probable hypothesis is not that one individual suddenly takes after an ancestor removed by some hundred generations, but rather that in each successive generation the character in question has been somehow lying latent and at last, under unknown favorable conditions, is developed.[17] With the barb pigeon (see Figure 1.2), for example, which very rarely produces a blue individual, there is probably a latent tendency in each generation to produce blue plumage.[18]

If all the species in the same genus are indeed descended from a common ancestor, as I claim, we might expect them to occasionally vary in an analogous manner, so that the different varieties of two or more species would occasionally resemble each other, or that a variety of one species would occasionally resemble a different distinct species in certain characteristics—after all, this other species would be, according to our view, only a well-marked and permanent variety. But characteristics that are exclusively due to analogous variation would probably be of an unimportant nature, since the preservation of all functionally important characteristics will have been fixed through natural selection, in accordance with the different habits of the different species. We might further expect that the various species within any one genus would occasionally exhibit reversions to long lost characteristics that represent those of the ancient ancestor of all current members of that genus. As, however, we do not know the common ancestors of *any* natural groups, we cannot distinguish such reversionary characteristics from those that are instead analogous. If, for instance, we did not know that the ancestral rock pigeon was not feather-footed (Figure 5.11A) or turn-crowned (Figure 5.11B), we could not know whether the

---

[17] Darwin is so close to the truth here! Surely a reading of Mendel's 1866 paper would have given him the mechanism he was looking for—the idea of discrete alleles that never change, except through mutation. If only one of Darwin's friends had sent him a copy of that paper…who knows what might have happened?

[18] Yes! Because the rare recessive allele for blue plumage never completely disappeared from the population.

(A)

(B)

Figure 5.11 (A) A feather-footed pigeon. (B) Turn-crowned pigeons are characterized by the feathers on the head projecting backwards.

appearance of such characteristics in our domestic breeds were reversions to an ancestral state or only analogous variations. Perhaps we would have inferred that the blue color was a case of reversion, based on the number of distinctive markings that are correlated with this tint, and that would probably not have all appeared together from simple variation. More especially we might have inferred reversion from the blue color and from the several marks appearing so often when differently colored breeds were crossed. Under nature, certainly, we cannot usually know which cases are reversions to formerly existing characteristics and which ones are new but analogous variations; even so, according to our theory, we can expect to find some of the varying offspring of a species assuming characteristics that are already present in at least some other members of the same group. And this is undoubtedly the case.

The difficulty in characterizing variable species is largely due to the varieties within that species mocking, as it were, other species in the same genus. A considerable list could also be given of forms intermediate between two other forms, which themselves can only doubtfully be ranked as species; and this shows, unless all these closely allied species are assumed to have been independently created, that some species have, by varying, assumed some of the characteristics of the others.

But the best evidence of analogous variation is afforded by organs or other body parts that are generally constant in character, but which occasionally vary so as to resemble, in some degree, the same part or organ in an allied species. I have collected a long list of such cases—including plant, insect, crustacean, reptile, bird, and mammalian examples; but here, as before, I lie under the great disadvantage of not being able to give them all. I can only repeat that such cases certainly occur, and seem to be very remarkable.

I will, however, give one curious and complex case, one that does not affect any important character but that occurs in several species of the same genus, partly under domestication and partly in nature. It is a case almost certainly of reversion to an ancestral state. I am talking here about donkeys and horses. The donkey (*Equus africanus asinus*, also commonly known as the "ass") (Figure 5.12A) sometimes has very distinct transverse bars on its legs, like those on the legs of the zebra. It has been said that these bars are plainest in the foals, and from the inquiries that I have made, I believe this to be true. They also have a stripe on the shoulder, sometimes a double stripe that varies considerably in both length and outline. A white donkey (but not an albino) has been described that has neither a spinal nor a shoulder stripe; these stripes are sometimes very obscure, or even completely lost, as is the case for dark-colored donkeys. The wild horse from the central Asian plains, the koulan, first documented by the German biologist Peter Simon Pallas, is said to have been seen with a double shoulder stripe. The English zoologist Mr. Edward Blyth, now in India, has seen a specimen of the onager (*Equus hemionus*)—the "Asiatic wild ass" of central Asia—with a distinct shoulder stripe, though it normally has none; and I have been informed by that expert

(A)   (B)

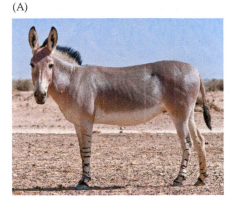 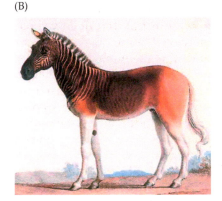

Figure 5.12 (A) A donkey (*Equus africanus asinus*); donkeys were first domesticated about 5,000 years ago. (B) The quagga (*Equus quagga quagga*) was a subspecies of the plains zebra. It became extinct in the wild by 1878, and the last zoo animal died in 1883.

on the horses of India, Colonel Skeffington Poole, that the foals of this species are generally striped on the legs, and faintly on the shoulder as well. The quagga (*Equus quagga quagga*) (Figure 5.12B) of South Africa, though so plainly barred like a zebra over its body, has no bars on its legs; but Dr. Asa Gray of Harvard University has illustrated one specimen with very distinct zebra-like bars on the animal's hocks, a region of the animal's hind limb.

In England, I have recorded cases of the spinal stripe in horses of the most distinct breeds, and of all colors. Transverse bars on the legs are fairly common among some breeds (duns and mouse duns[19]) and I have also seen this in one instance in a chestnut; duns sometimes also have a faint shoulder stripe, and I have seen a trace of one in a bay horse as well. The term dun includes a large range of coat colors, from one between brown and black to something closely approaching cream-colored. My son made a careful sketch for me of a dun Belgian cart horse with a double stripe on each shoulder and with leg stripes as well, and I myself have seen a dun Devonshire pony with three parallel stripes on each shoulder. A small dun Welsh pony with the same three parallel stripes has also been carefully described to me.

In northwest India, the Kathiawari breed of horses is so often striped that, according to Colonel Poole, who examined this breed for the Indian government, a horse without stripes is not considered as purebred. The spine is always striped; the legs generally show bars; they commonly have a shoulder stripe, which is sometimes double and sometimes triple; and even the side of the face is sometimes striped. The stripes are often plainest on the foals, and sometimes they disappear in older horses. Colonel Poole has seen both gray

---

[19] Dun horses have a lighter hair coloration than normal. Mouse dun refers to a dark brownish gray color. We now know that these color patterns are caused by a simple dominant gene called "dun," which affects the deposition of both red and black pigment equally.

and bay Kathiawari horses striped when first foaled. I also have good cause to suspect, from information given to me by Mr. W. W. Edwards, that with the English racehorse, the spinal stripe is also much commoner in foals than in full-grown animals. I myself have recently bred a foal from a bay mare (the offspring of a Turkoman horse and a Flemish mare) mated with a bay English racehorse. When this foal was a week old, it had many very narrow, dark, zebra-like bars on its hind quarters and on its forehead, and its legs were feebly striped; all the stripes soon disappeared completely as the horse grew. Without here entering still further details, let me just state that I have collected cases of leg and shoulder stripes in horses of very different breeds in various countries from Britain to Eastern China, and from Norway in the north to the Malay Archipelago in the south. In all parts of the world, these stripes occur most often in duns and mouse duns.

Charles Hamilton Smith, who has written on this subject, believes that the several breeds of horse are descended from several different aboriginal species, one of which, the dun, was striped, and that the above-described appearances are all due to ancient crosses with the dun stock. But we may safely reject this view. It is highly improbable that the heavy Belgian cart horse, Welsh ponies, cobs, the lanky Kathiawari race, and so forth, inhabiting the most distant parts of the world, should all have been crossed with one supposed aboriginal stock.

Now let us turn to the effects of cross-breeding among the several species in the horse genus (*Equus*). Rollins asserts that the common mule—the offspring of matings between two different species: a male donkey and a female horse—is particularly likely to have bars on its legs; indeed, according to the English zoologist Mr. Philip Henry Gosse, in certain parts of the United States about nine out of ten mules have such striped legs. I once saw a mule with its legs so much striped that anyone might have thought it was a hybrid zebra; and Mr. William Charles Martin's excellent treatise (*History of the Horse*, 1845) includes a figure of another similar mule. In four colored drawings that I have seen of hybrids from matings between donkeys and zebras, the legs were much more plainly barred than the rest of the body; and in one of them there was a double shoulder stripe. Lastly, and this is another most remarkable case, a hybrid from matings between the donkey and the onager has been illustrated by Dr. Gray; this hybrid had three short shoulder stripes, like those on the dun Devonshire and Welsh ponies, and also had some zebra-like stripes on the sides of its face, even though the donkey only occasionally has stripes on its legs and the onager has none; it doesn't even have a shoulder stripe. With respect to the zebra-like stripes on the hybrid's face, I was so convinced that stripes of color never appear from what is commonly called chance, that I was led to ask Colonel Poole whether such face stripes ever occurred in the eminently striped Kathiawari (also called Kattywar) breed of horses; as we have seen, the answer was yes.

How on earth can we explain all of these facts? We see several distinct species of the horse genus becoming, by simple variation, striped on the legs like a zebra, or striped on the shoulders like a donkey. In the horse we see this tendency very strongly whenever a dun tint appears—a tint that approaches that of the general coloring of the other species of the genus. The appearance of the stripes is not accompanied by any change of form or by any other new characteristic. We see this same tendency to become striped most strongly displayed in hybrids from between several of the most distinct species.

To understand these facts, let us consider the several breeds of pigeons as a parallel case: all of the breeds are descended from a pigeon of a bluish color, with certain bars and other distinctive marks (the rock pigeon; see Figure 1.3D); and when any breed assumes by simple variation a bluish tint, these bars and other marks invariably reappear, and without any other change of form or character. When the oldest and truest different pigeon breeds of various colors are crossed, we see a strong tendency for the blue tint and bars and marks to reappear in the mixed offspring. As I have previously stated, the most probable hypothesis to account for the reappearance of these very ancient characteristics is that there is a tendency in the young of each successive generation to produce the long-lost character, and that this tendency, from unknown causes, sometimes prevails. And we have just seen that in several species of the horse genus, the stripes are either plainer or appear more commonly in the young than in the old. If we were to think of the various pigeon breeds, some of which have bred true for centuries, as separate species, then the case would be exactly parallel with that of species in the horse genus: we see striping patterns periodically among members of the genus *Equus* because the common ancient ancestor was striped! I am confident that were we to look back thousands on thousands of generations, we would see in the common parent of the donkey, the onager, the quagga, the zebra, and our domestic horse, an animal striped like a zebra but perhaps otherwise very differently constructed. **All of these animals must have descended from the same striped ancestor long, long ago.**

Anyone who believes that each equine species was independently created will, I presume, assert that each species has been created with a tendency to vary in this particular manner, both under nature and under domestication, so as often to become spontaneously striped like the other species of the genus; and that each has been created with a strong tendency, when crossed with species inhabiting distant quarters of the world, to produce hybrids resembling in their stripes, not their own parents, but other species of the genus. To admit this view, it seems to me, is to reject a real cause for an unreal cause—or at least an unknown one. It makes the works of God a mere mockery and deception; I would almost as soon believe the medieval assertion that fossil shells had never actually contained living animals, but had instead been created in stone so as to mock the shells now living on the seashore.

Summary

Our ignorance of the laws of variation is profound. Not in one case out of a hundred can we pretend to explain why this or that part has varied as it has. **But whenever we have the means of making a comparison, the lesser differences between varieties of the same species and the greater differences between species of the same genus all seem to have been produced through the same laws.** Here are some of the things we think we know. Changed environmental conditions may somehow induce variability, habit and flexibility (plasticity) may produce constitutional peculiarities, and use and disuse may strengthen or weaken organs, respectively. Symmetrical parts tend to vary in the same ways and to the same degree, and some other parts are somehow related to each other so that variations in one part cause those others parts to become modified as well. Modifications in hard parts and in external parts can sometimes affect the form of softer and internal parts. When one part is largely developed, perhaps it draws nourishment from adjoining parts; but every part of the structure that can be saved without detriment will be saved. Changes of structure at an early age may affects parts developed later in life; and many such cases of "correlated variation," the mechanism behind which we are unable to understand, undoubtedly do occur.

In addition, multiple parts are especially variable both in number and in structure, perhaps because such parts have not been closely specialized for any particular function, so that their modifications have not been closely checked by natural selection. It follows, probably from this same cause, that so-called lower organisms are more variable than those standing higher on the scale, and which have their whole organization more specialized. Rudimentary organs, such as the eyes in some cave animals, from being useless, are not regulated by natural selection and thus are free to vary; and vary they do. Species-defining characteristics (i.e., specific characteristics) are more variable than those that define genera (generic characteristics), or those that have long been inherited and have not differed from this same period.

In this chapter I have sometimes referred to special parts or organs being still variable because they have recently varied and thus come to differ; but we have also seen in Chapter 2 that the same principle applies to the whole individual: for in a region where many species of a particular genus are found—that is, where there has been much former variation and differentiation, or where the production of new specific forms has been actively at work—in that district and amongst those species, we now find, on average, the most varieties.

Secondary sexual characteristics are highly variable, and such characteristics differ much among species in the same group. Variability in the same anatomical features has generally been taken advantage of by selection in giving secondary sexual differences to the two sexes of the same species as well as specific differences to various species within the same genus.

Any part or organ developed to an extraordinary size or in an extraordinary manner, in comparison with the same part or organ in allied species within the genus, must have gone through an extraordinary amount of modification since the genus arose. Thus we can understand why it should often still be variable in a much higher degree than other parts, for variation is a long-continuing and slow process, and natural selection will in such cases not as yet have had time to overcome the tendency to further variability and reversion to a less modified state. But when a species with any extraordinarily developed organ has given rise to many modified descendants—which in my view must be a very slow process, requiring a great lapse of time—in this case, natural selection has succeeded in giving a fixed character to the organ, in however extraordinary a manner it may have been developed.

Species inheriting nearly the same constitution from a common parent and exposed to similar influences naturally tend to present analogous variations, or these same species may occasionally revert to some of the characteristics of their ancient ancestors. Novel modifications, however they arise, will add to the beautiful and harmonious diversity of nature.

Whatever the cause may be of each slight difference between offspring and their parents—and a cause for each *must* exist!—we have reason to believe that it is the steady accumulation of beneficial differences over a great many generations that has given rise to all the more important modifications of structure in relation to the habits of each species.

Key Issues to Talk and Write About

1. Explain Darwin's argument that the occasional appearance of a blue color and certain distinctive markings when different varieties of pigeons are bred together tells us much about the ancient ancestor of the modern horse and donkey.

2. Based on the arguments given in this chapter, which of the following do you think Darwin makes the most convincing case for, and why does that argument seem most convincing to you?

   a. Use and disuse of parts

   b. Correlated variation

   c. Compensation

3. What does Darwin mean by "analogous variation"? Give an example.

4. Find out two interesting things about one of the people that Darwin mentions in this chapter. Choose from the following:

   Thomas Vernon Wollaston   Johann von Goethe
   Richard Owen   George Robert Waterhouse
   John Obadiah Westwood   Edward Blyth
   Louis Agassiz   Peter Simon Pallas

Benjamin Silliman	Charles Hamilton Smith
Hewett Cottrell Watson	Augustin de Candolle
Asa Gray	Christian Konrad Sprengel
Isidore Geoffroy Saint-Hilaire	Étienne Geoffroy Saint-Hilaire

5. Explain what Darwin means by the term "secondary sexual characteristics."

6. Following the guidelines given in Chapter 1 (see page 28, Key Issue 10), choose one of the following paragraphs from Chapter 5 and write an informative but standalone one-sentence summary of that paragraph. Choose either of the following two paragraphs:

   a. Page 123, the paragraph that begins, "I also suspect that some of the presumed cases of compensation that have been brought forward…" or

   b. Page 135, the paragraph that begins, "In northwest India, the Kathiawari breed of horses is so often striped that…"

7. Try your hand at revising the following sentences, to make them clearer and more concise:

   a. The same number of joints in the tarsi is a character common to very large groups of beetles, but in the Engidae the number varies greatly.

   b. They also have a stripe on the shoulder that is very variable in length and outline, and is sometimes double.

Online Resources *available at* sites.sinauer.com/readabledarwin

Video

5.1 Differences in appearance and behavior in male and female pheasants

Links

5.1 Sexual dimorphism in a number of animals, including birds
http://www.bbc.co.uk/nature/adaptations/Sexual_dimorphism#intro

5.2 Stripes appearing on donkeys, mules, and horses
http://www.brindlehorses.com/stripedhos/patton-article/patton.htm

5.3 600 horse breeds, including those mentioned by Darwin
http://www.theequinest.com/breeds/

(*Note: Web addresses may change. Go to sites.sinauer.com/readabledarwin for up-to-date links.*)

Bibliography

Kirby, W. F. and W. Spence. 1815–26. *An Introduction to Entomology: Or Elements of the Natural History of Insects.* London.

Martin, W. C. 1845. *History of the Horse.* London.

6

Difficulties with the Theory

If modern animals and plants have evolved to their present condition slowly over long periods of time, why don't we now see clear intermediate forms between all of the existing species? And how could incredibly complicated organs like the vertebrate eye possibly have developed gradually by small steps? If the intermediate stages weren't functional (what good is half an eye?), how could natural selection possibly account for the evolution of such structures to their present state of great complexity? And how could natural selection be at all involved in causing the development of body parts that appear to play no important role in an organism's life? And how can anyone argue that flowers, diatoms, and other such beautiful organisms weren't created especially for us to admire? Darwin answers all of these questions patiently, logically, and brilliantly, and with one marvelous example after another.

Long before readers have arrived at this part of my work, a crowd of difficulties will have occurred to them. Some of the difficulties are so serious that to this day I can hardly think of them without being in some degree staggered. But to the best of my judgment, most of the imagined difficulties pose no real problems, and those that are real are not, I think, fatal to the theory.

Each of these difficulties and objections may be placed under one of the following four headings:

1. First, if all species have descended from other species by very small steps, then why do we not everywhere see innumerable transitional forms between species? Why is not all of nature in confusion, instead of each species being, as we see them, so well defined?

2. Secondly, is it really possible that an animal having, for instance, the structure and habits of a bat, could have been formed by the modification of some other animal with very different habits and a very different structure? And can we honestly believe that natural selection could produce, on the one hand, an organ of trifling importance such

as a giraffe's tail, which serves merely as a flyswatter, and, on the other hand, an organ so wonderfully complex as the eye?

3. Thirdly, can instincts also be acquired and modified through natural selection? What about, for example, the instinct that leads bees to make beehives, whose inner cells are so complex as to practically anticipate the discoveries of profound mathematicians?

4. Fourthly, how can we account for the fact that when members of different species are mated together, they either produce no offspring or produce offspring that are themselves sterile, whereas when distinct varieties of the same species are crossed, they have no trouble producing viable offspring?

I will discuss the first two points in this chapter, and the other points in succeeding chapters.

On the Absence or Rarity of Transitional Varieties

As natural selection acts solely by preserving advantageous modifications, each new superior form will tend, in the midst of intense competition or predation pressure, to take the place of, and finally to exterminate, its own less improved predecessors. Extinction and natural selection go hand in hand. Thus, if we look at each species as being descended from some unknown ancestral form, both the parent and all of the transitional varieties will generally have been exterminated by the very process of forming and gradually perfecting the new form.

But there is a problem: if innumerable transitional forms have existed in former times, why do we not now find them embedded in countless numbers in the earth's crust? I will discuss this question in detail later, in the chapter On the Imperfection of the Geological Record (Chapter 10).[1] Here I will simply say that I believe the answer lies mainly in the geological record being incomparably less perfect than is generally supposed. Although the earth's crust is indeed a vast museum, the natural collections have been imperfectly preserved, and only at long intervals of time. I will talk more about this later.

It may also be argued that when several closely related species inhabit the same territory at the same time, we surely ought to find at the present time many transitional forms filling in the gaps between them. Let us a take a simple case: in traveling from north to south over a continent, we generally encounter at successive intervals a series of closely allied or representative species, **each evidently filling nearly the same ecological niche where it lives**. These representative species often meet and overlap, and as the one becomes rarer and rarer, the other becomes more and more frequent, until

---

[1] Darwin's cross-references to chapters beyond Chapter 8 have been retained in this volume so readers can refer to the original *The Origin of Species*, Sixth Edition.

the one fully replaces the other as we continue our travels. But if we compare these species where they intermingle, they show no intermediate traits: they are generally as absolutely distinct from each other in every detail of structure as are specimens taken from the centers of their distribution. How can we explain this?

According to my theory of natural selection, these allied species are descended from a common ancestor, so that during the process of evolution each has gradually become adapted to the conditions of life in its own region, and has supplanted and exterminated its original ancestral form and all the transitional varieties between its past and present states. Thus we should not expect to see numerous transitional varieties in each region now, though they must have existed there at one time in the past, and may in fact be embedded there still as fossils. But in the regions between where we now find the two related species, having intermediate conditions of life, why do we not now find closely linking intermediate varieties? For a long time this problem quite confounded me. But I think I can now in large part explain it.

In the first place, just because an area of land is now continuous, we should be extremely cautious in assuming that it has always been so. Geology leads us to believe that most continents have in fact been broken up into a number of separate islands even within the past several million years. On such isolated islands, distinct species might have gradually evolved without the possibility of intermediate varieties existing in intermediate zones.

But I will pass over this way of escaping from the difficulty; although I do not doubt that the *formerly* isolated condition of lands that are now continuous has played an important part in forming new species—most especially with freely crossing and widely wandering animals—I believe that many perfectly defined species have also formed within strictly continuous areas.

Here is my reasoning. In looking at species as they are now distributed over a wide area, we generally find that the individuals of each species are fairly numerous over a large territory, and then become somewhat abruptly rarer and rarer at the extremes of their range, finally disappearing altogether. Thus the neutral territory *between* two representative species is generally small and narrow compared with the territory dominated by each. We see the same thing when we climb mountains: sometimes it is quite remarkable how abruptly a common alpine tree species disappears as we continue to climb upward, as the Swiss botanist Alphonse de Candolle has observed. The same sharp boundaries between species have been noticed by the British naturalist Professor Edward Forbes in dredging up animals from the depths of the sea.

To those who look at climate and the other physical conditions of life as the all-important elements controlling the distributions of animals and plants, these facts ought to cause surprise, as climate, elevation, and water depth graduate away almost imperceptibly. But distributions are not controlled only by physical conditions. Remember that almost every species, even where it is the most abundant, would increase immensely in numbers were it not for the impact of other species; the members of nearly all species either prey on other species or serve as prey for others—i.e., each living organism is either directly or indirectly related in the most important manner to other living organisms. **Thus we see that the area occupied by any given species in any given country by no means depends exclusively on barely perceptible, gradually changing physical conditions, but rather in large part on the presence of the other species on which it feeds, or by which it is destroyed, or with which it competes for food or space**. And as all of these other species are already well-defined objects, not blending into one another by small, barely perceptible graduations, the range of any one species, depending as it does on the range of other species, will tend to be sharply defined.

Moreover, on the edges of its range, where each species exists in smaller numbers, the species will be extremely liable to utter extermination during fluctuations in the number of its enemies or of its prey, or in climate. The geographical range of the species will then become even more sharply defined.

We can probably apply this same rule—that two or more allied or representative species occupying a continuous area generally have wide individual ranges with a much narrower neutral territory between them, and become rarer and rarer within that narrow zone of overlap—to the distribution of *varieties within* a given species. Thus, if we take a varying species inhabiting a very large area, we shall have to adapt two varieties to two large areas, and a third variety to a narrow zone between them. The intermediate variety in that narrow intermediate zone must exist in smaller numbers, because it inhabits a narrow and smaller area. As far as I can tell, this rule indeed holds good with varieties in nature. I have in fact met with striking examples of this rule in the case of individuals intermediate between well-marked varieties in the barnacle genus *Balanus* (see Figure 5.8). Furthermore, it would seem from information given to me by the botanists Mr. Hewett Watson and Dr. Asa Gray, and the entomologist Mr. Thomas Vernon Wollaston, that when varieties intermediate between two other forms occur, they are indeed generally found in much smaller numbers than the forms that they connect.

If we trust these facts and inferences and thus conclude that intermediate varieties linking two other varieties together generally have existed in smaller numbers than the forms that they connect, then we can understand why intermediate varieties should not last very long in nature—why, in fact, they should generally be exterminated and disappear, and do so sooner than the forms that they originally linked together. For any form existing in fewer numbers would, as already remarked, run a greater chance of being exterminated than one existing in large numbers. In this particular case the intermediate form would be readily susceptible to the pressures imposed by the closely related forms existing on both sides of it. But the really important point here is that during the gradual process of further modification, by which two varieties will presumably develop into two distinct species, the two forms that exist in larger numbers, from inhabiting larger areas, will have a great advantage over the intermediate variety that exists in small numbers and lives in a narrow zone in between the other two. **Forms existing in larger numbers will always have a better chance of presenting further favorable variations for natural selection to work with than will the rarer forms that are fewer in number.**

In the never-ending race for life, then, the more common forms will always tend to beat and supplant the less common forms, for the less common forms will be modified and improved over time much more slowly, simply because there are fewer individuals involved. I believe it is this very same principle that accounts for the most common species in each country presenting on average more well-marked varieties than do the rarer species (see Chapter 2). Let me illustrate: Suppose we have three varieties of sheep: the first is adapted to an extensive mountainous region; the second is adapted to a comparatively narrow, hilly tract of land; and the third is adapted to the wide plains at the base of the mountain. Suppose, too, that the sheep herders in these areas are all trying with equal steadiness and skill to improve their stocks by artificial selection. The great sheep holders on the mountains or on the plains will surely improve their breeds more quickly than the small holders on the intermediate, narrow, hilly tract. Consequently, either the improved mountain breed or the improved plains breed will soon replace the less improved hill breed. Thus, the two breeds that originally existed in greater numbers will now come into close contact with each other, without the interposition of the now-supplanted, intermediate hill variety.

To summarize, I believe that all species come to be reasonably well-defined, and to never present an inextricable chaos of varying and inter-mediate links, for several reasons. First, because variation is a slow process, new varieties can form only very slowly. Remember, natural selection can do nothing until favorable individual differences or variations occur, and until a niche in some particular part of the country can be better filled by some modification of some one or more of its inhabitants. The advent of new niches in an area will depend on slow changes of climate or on the oc-casional introduction of new inhabitants (e.g., through immigration), and probably—and even more importantly—on some of the old inhabitants becoming slowly modified themselves over many generations. The new forms thus produced, along with the old ones, would then act on and react to each other, so that, in any one region and at any one time, we ought to see only a few species presenting slight modifications of structure that are in some degree permanent. And this is in fact what we see.

Secondly, areas that are now continuous must often have existed as iso-lated pockets in which many forms may have separately become sufficiently distinct to rank as representative species of the area. This is especially likely for animals that congregate for mating and that wander far and wide over the entire area. In this case, varieties that were intermediate between the several representative species and their common parent must formerly have existed within each isolated portion of the land...but those intermediate varieties will have been supplanted and exterminated over time by natural selection, and thus will no longer be living.

Thirdly, when two or more varieties have been formed in different parts of a strictly continuous area, intermediate varieties will probably have formed in the zones between them; but those intermediate varieties will generally have been short-lived. Why? Because the intermediate varieties will, for reasons that I have already discussed, exist in the intermediate zones in fewer numbers than the varieties that they connect. For this reason alone, the intermediate varieties will be readily vulnerable to accidental extermination. Moreover, organisms that exist in greater numbers will, on average, present more varieties and are consequently more likely to become further improved over time through natural selection, thereby gaining further advantages over any intermediates. Thus, over time, the intermediates will almost certainly be beaten and supplanted by the forms that they once connected.[2]

Lastly, looking not to any one time but to all time, if my theory is correct then countless intermediate varieties that linked all the species within any particular group close together must assuredly have once existed. However, the process of natural selection tends to constantly exterminate both the ancestral forms and the intermediate links. Thus, evidence of their former existence will only be found among their fossil remains; however, these remains are preserved in an extremely imperfect and intermittent record, as I shall attempt to show in Chapter 10.

On the Origin and Transitions of Organic Beings with Peculiar Habits and Structure

Opponents of my views have asked, how could a terrestrial meat-eating (i.e., carnivorous) animal have been converted into one with aquatic habits? How could such an animal have survived in its transitional state? Well, a number of existing carnivorous animals do in fact show very clear intermediate grades from strictly terrestrial to aquatic habits; as each experiences a severe struggle for life, it is clear that each must be quite well adapted to its place in nature.

Look, for example, at the American mink (*Neovison vison*) of North America (Figure 6.1A). It has webbed feet, like a duck, but resembles an otter in its fur, short legs, and in the form of its tail. During the summer this

---

[2] We now also have evidence that there can be selection against interbreeding between individuals belonging to adjacent varieties. Such selection will discourage the formation of intermediate varieties in the first place.

(A)

(B)

Figure 6.1 (A) The American mink (*Neovison vison*).
(B) Flying squirrel (*Glaucomys volans*). (C) The flying lemur
(colugo) (*Galeopterus variegatus*) from the rainforests of
Malaysia.

(C)

animal dives for and feeds on fish, but during
the long winter it leaves the frozen water and
feeds, like other polecats, on mice and other
land animals. If a different case had been taken,
and I had been asked how an insect-eating
four-legged animal could possibly have been
converted into a flying bat, the question would
have been far more difficult to answer. Yet I
think that such difficulties have little weight.

Here, as on other occasions, I lie under a
heavy disadvantage: out of the great many strik-
ing cases that I have collected, I have space here to give only one or two
examples of transitional habits and structures found in related species, and
of diversified habits among members of the same species.

Look at the family of squirrels, for instance, which includes at least 200
distinct species. Here we have the finest gradations, from some animals with
their tails only slightly flattened to others with, as the naturalist Sir John
Richardson has noted, the posterior part of their bodies rather wide and
with the skin on their flanks rather full, and finally to the so-called flying
squirrels, which have their limbs—and even the base of their tails—united
by a broad expanse of skin that serves as a parachute and allows them to
glide through the air to an astonishing distance from tree to tree (Figure
6.1B). We cannot doubt that each modification is of some use to each kind
of squirrel in its own habitat, by enabling it to escape birds or other beasts
of prey, or to collect food more quickly, or to lessen the danger from occasional falls. But it does not follow from this fact that the structure of each
squirrel is the *best* that it could possibly be under all possible conditions.
Let the climate and vegetation change over time, or let other competing
rodents or new beasts of prey immigrate into the squirrel's territory, or let
old ones within the territory become gradually modified, and we would

expect at least some of the squirrels to decrease in numbers or become ex-terminated…unless they also gradually became modified and improved in structure in a corresponding manner in subsequent generations. Therefore I can see no difficulty, especially under changing environmental conditions, in the continued preservation of individuals with fuller and fuller flank membranes, with each modification being useful and each therefore being propagated to future generations, until by accumulated effects of this process of natural selection over long periods of time, a perfect so-called flying squirrel was produced.

Now let's consider the Southeast Asian flying mammals known as co-lugos (Figure 6.1C), a group formerly ranked amongst the bats but that is now believed to belong in a separate order.[3] An extremely wide flank membrane stretches from the corners of the jaw to the tail of these animals, and includes limbs with elongated fingers. This flank membrane is furnished with an extensor muscle. Although no graduated links of structure fitted for gliding through the air now connect the colugos with any other members of the order, there is no difficulty in supposing that such links existed in the past, or that each was developed in the same manner as with the less perfectly gliding squirrels, with each modification of structure having been useful to its possessor. Nor can I see any insuperable difficulty in further believing that the membrane-connected fingers and forearm of the colugos might have been greatly lengthened by natural selection and this, as far as the organs of flight are concerned, would have converted the animal into a bat. In certain bats the wing membrane extends from the top of the shoulder to the tail and includes the hind legs; here, perhaps, we see traces of an apparatus originally suited for gliding through the air, rather than for flight.

If about a dozen genera of birds had become extinct before our time, who would have ever thought that some birds might have once used their wings solely as flappers, like the flightless steamer duck (see Figure 5.1) does today; or as fins in the water and as front legs on land, like the penguin; or as sails, like the ostrich; or functionally for no purpose at all, as in the kiwi (Figure 6.2). Yet the structure of each of these birds is good for it under the conditions of life in which it lives, for each has to live by a struggle. But it is not necessarily the *best* structure possible under all possible conditions. Please do not infer from my remarks that any of the grades of wing structure here alluded to indicate the steps by which birds actually acquired their perfect power of flight; but they do serve to show what

Figure 6.2 An immature and adult male kiwi (*Apteryx haastii*).

---

[3] This is now known as the order Dermoptera. Only two colugo species exist today, both still restricted to Southeast Asia.

diversified means of transition from state to state are at least possible as animals adapt to different conditions.

Let us continue this argument. Seeing that some members of such primarily water-breathing groups as the Crustacea (a group that includes the crabs and hermit crabs) and the Mollusca (the phylum that includes the snails) are adapted instead to life on the land, and seeing that we have both flying birds and flying mammals, and flying insects of the most diversified types, and that we formerly had flying reptiles, it is conceivable that flying fish (Figure 6.3), which now glide far through the air, slightly rising and turning by the aid of their fluttering fins, might have become

Figure 6.3 Flying fish shortly after take-off.

modified by now into perfectly winged animals. If this had occurred, who would have ever imagined that in an early transitional state they had inhabited the open ocean and had used their incipient organs of flight exclusively, as far as we know, to escape being devoured by other fish?

When we see any structure highly perfected for any particular habit, such as the wings of a bird for flight, we should bear in mind that animals displaying early transitional grades of the structure will seldom have survived to the present day; rather, they will have been supplanted by their successors, which were gradually rendered more perfectly adapted for their lifestyle through natural selection over a great many generations. Furthermore, we may conclude that transitional states between structures suited for very different habits of life will rarely, when first developed, have appeared in great numbers and under many subordinate forms.

Thus, to return to our imaginary illustration of the flying fish, it seems unlikely that fishes capable of true flight would ever develop under many subordinate forms—for taking prey of many kinds in many ways, for example, on the land and in the water—until their organs of flight had come to a high degree of perfection, so as to have given them a decided advantage over other animals in the battle for life. Thus the chance of discovering species with transitional grades of structure in fossils will always be less than in the case of species with fully developed structures, simply because the transitional forms will have existed in smaller numbers.

I will now give a few examples of changed and diversified habits in individuals of the same species. In either case, it would be easy for natural selection to adapt the structure of the animal to its changed habits, or exclusively to just one of its several habits. It is, however, difficult to decide whether habits generally change first and structure afterward, or whether slight modifications of structure lead to changed habits; both probably occur

Figure 6.4 A great kiskadee (*Pitangus sulphuratus*) standing on a tree branch.

Figure 6.5 The great tit (*Parus major*).

Figure 6.6 The common treecreeper (*Certhia familiaris*).

Figure 6.7 A black bear (*Ursus americanus*).

almost simultaneously in many cases. But really, this problem is immaterial for us.

For cases of changed habits, I need only remind readers of the many British insects that now feed on nonnative plants, or even feed exclusively on artificial substances. Of diversified habits, I could give innumerable examples. I have, for example, often watched a bird in South America known as the great kiskadee (*Pitangus sulphuratus*, also called the tyrant flycatcher) (Figure 6.4) hovering over one spot and then proceeding to another, just like a kestrel does, and at other times standing stationary at the water's edge and then dashing into it, like a kingfisher behaves towards a fish. In our own country, the great tit (*Parus major*) (Figure 6.5) may be seen climbing branches, almost like a treecreeper (Figure 6.6); but then sometimes it kills small birds by blows on the head, like a shrike. And I have many times seen and heard it hammering away at the seeds of the yew tree on a branch, and thus breaking them like a nuthatch does. In North America, the black bear (Figure 6.7) was seen by the English explorer, fur trader, and naturalist Samuel Hearne swimming for hours with its mouth open wide, thus catching insects in the water almost like a baleen whale catches krill.

As we sometimes see individuals following habits different from those normal for their species and even for other species in the same genus, we might expect that such individuals would occasionally give rise to new species having such anomalous habits, and with their structure either slightly or considerably modified from that of their normal fellows.

Such instances do indeed occur in nature. Can a more striking instance of adaptation be given than that of a woodpecker, for climbing trees and seizing insects in the chinks of the bark? Yet in North America, there are some woodpeckers that feed largely on fruit, and others with elongated wings that they use to chase insects in flight. And on the plains of La Plata, Argentina, where hardly a trees grows, there is a woodpecker species (*Colaptes campestris*) (Figure 6.8) that has two toes before and two behind, a long pointed tongue, pointed tail feathers sufficiently stiff to support the bird in a vertical position on a post (but not as stiff as those of typical woodpeckers), and a straight, strong beak. The beak, however, is not as straight or as strong as in typical woodpeckers, but it is strong enough to bore into wood. This animal is, in all the essential parts of its structure, clearly a woodpecker. Even in such trifling characters as its coloring, the harsh tone of its voice, and its undulatory flight pattern, its close blood relationship to our common woodpecker is plainly declared. Yet, as I can assert from my own observations as well as those of the South American explorer and naturalist Félix de Azara, in certain large districts this bird does not climb trees; indeed, it makes its nest in holes in the banks of rivers! In certain other districts, this very same woodpecker frequents trees and bores holes in the trunk for its nest, as described by the ornithologist Mr. William Henry Hudson. I may mention as another illustration of the varied habits found within this genus, that a Mexican *Colaptes* species has been said by the Swiss zoologist, Henri Louis Frédéric de Saussure to bore holes into hard wood in order to lay up a store of acorns. The habits and lifestyles of woodpeckers are indeed diverse.

Petrels (Figure 6.9A, B) offer another example of diverse habits. They are generally the most aerial and oceanic of birds,

Figure 6.8 The South American woodpecker, Pica-pau-do-campo (*Colaptes campestris*).

(A)

(B)

(C)

Figure 6.9 (A) The northern giant petrel (*Macronectes halli*). (B) A diving-petrel (*Pelecanoides urinatrix*). (C) A pied-billed grebe (*Podilymbus podiceps*). These are expert divers: "part bird, part submarine." They are common in freshwater marshes, sluggish rivers, and estuaries.

Figure 6.10 A Eurasian dipper (was waterouzel in Darwin's day), genus *Cinclus*.

but in the quiet waters off Tierra del Fuego, at the southern tip of South America, the petrel, in its general habits, in its astonishing power of diving, in its manner of swimming and of flying when made to take flight, would be mistaken by anyone for an auk or a grebe (Figure 6.9C); it is essentially a petrel, but with many of its structures profoundly modified over time to match its new habits of life. Compared with that bird, the woodpecker of La Plata has had its structure modified only slightly. In the case of dippers (members of the genus *Cinclus*) (Figure 6.10), even the most careful observer would never suspect its subaquatic habits merely by examining its dead body. Yet this bird, which is allied with the thrush family, subsists by diving, using its wings to move under the water and grasping stones with its feet.

Lastly, all of the hymenopterous[4] insects (class Insecta, order Hymenoptera) are terrestrial except for members of a single genus, *Proctotrupes*, which the biologist Sir John Lubbock has discovered to be aquatic in its habits. It often enters the water and dives about by the use of its wings rather than its legs, and remains as long as 43 hours beneath the surface before coming up again for air. And yet despite its abnormal habits, it exhibits no corresponding modifications in structure.

He who believes that each being has been created as we now see it must occasionally have felt surprise when meeting an animal having habits not in agreement with its structure. What can be plainer than that the webbed feet of ducks and geese are formed for swimming? And yet there are web-footed upland geese that rarely go near the water. And, as described by the French-American ornithologist John James Audubon, the frigate bird (Figure 6.11A) alights on the surface of the ocean using its toes, all four of which are webbed. On the other hand, grebes and coots are eminently aquatic, even though their feet are not webbed: their toes are only bordered by membrane. What seems plainer than that the long toes found in wading birds, which are not furnished with any membrane between the toes, are formed for walking over swamps and floating plants? And yet the purple gallinule (Figure 6.11B), the moorhen (Figure 6.11C), and corncrake (Figure 6.11D) are members of a single family (Rallidae), even though the moorhen is nearly as aquatic as the coot, and the corncrake is nearly as terrestrial as the quail or partridge. In such cases—and many others could be given—habits have changed without a corresponding change of structure. The webbed feet of the upland goose may be said to have become almost

---

[4] Members of the insect order Hymenoptera, which includes at least 150,000 species of bees, ants, sawflies, and wasps. In comparison, there are fewer than 5,500 mammal species.

(A)

(B)

Figure 6.11 (A) A male great frigatebird (*Fregata minor*). (B) Purple gallinule, water walker (*Porphyrio martinica*). (C) Moorhen (*Gallinula chloropus*). (D) Corncrake (*Crex crex*).

(C)

rudimentary in function, though not in structure. In the frigate bird, the deeply scooped membrane between the toes shows that that structure has begun to change.

He who believes in separate and innumerable acts of creation may say that in all of these cases it has pleased the Creator to cause a being of one type to replace a being of another type. But this seems to me to be only restating the facts in dignified language. **He who believes in the struggle for existence and in the principle of natural selection will acknowledge that every organic being is constantly endeavoring to increase in numbers, and that if one individual varies in a useful way ever so little, either in habits or in structure, and in doing so gains an advantage over some other inhabitant of the same region, it will seize on the place of that inhabitant, however different**

(D)

that may be from its own home. Such a person will not be surprised to learn of geese and frigate birds living on the dry land and rarely alighting on the water, despite their having webbed feet; or that there are long-toed corncrakes living in meadows instead of swamps; or that there are woodpeckers living where hardly a tree grows; or that there are diving thrushes, and diving hymenopteran insects, and birds such as petrels that live always at sea (except to breed).

Organs of Extreme Perfection and Complication

To suppose that the eye, with all its inimitable contrivances for adjusting the focus to different distances, for admitting different amounts of light, and for correcting spherical and chromatic aberration, could have been formed by natural selection, seems, I freely confess, absurd in the highest degree. But when it was first said that the Sun stood still and that the world revolved around it, the common sense of mankind mistakenly declared that doctrine to be false. It seemed obvious that the Sun revolved around the Earth. As every philosopher knows, however, the old Latin saying of *vox populi vox Dei* ("The voice of the people is the voice of God") cannot be trusted in science. Reason tells me the following:

- That if we can show the existence of numerous graduations from a simple and imperfect eye to one that is complex and perfect, with each grade being useful to its possessor, as is certainly the case; and

- if further, the eye's characteristics can vary among individuals and that those variations are inherited, as is also certainly the case; and

- if such variations should be useful to an animal under changing conditions of life,

then the difficulty of believing that a perfect and complex eye could eventually be formed by natural selection, though perhaps almost impossible for our imaginations to grasp, should not be considered to doom the theory. How a nerve comes to be sensitive to light hardly concerns us any more than how life itself originated, but as some of the lowest organisms can perceive light even though no nerves can be detected, it is clearly possible that certain sensitive elements in their cytoplasm should have eventually become aggregated and developed into nerves endowed with this special sensibility.

In searching for the stages through which any particular organ in any particular species has been perfected, we should look exclusively to its direct ancestors. But this is scarcely ever possible, and so we are forced to look to other species and genera belonging to the same group—that is, to the collateral descendants, the nieces and cousins descended from the same ancestor—in order to see what gradual stages are possible, hoping for the chance of finding some gradations that have been transmitted in an unaltered or little-altered condition. The state of the same organ in distinct classes of organisms may incidentally throw light on the steps by which it has been perfected. Let us take this approach in trying to understand the evolution of eyes.

The simplest organ that can be called an eye consists merely of an optic nerve surrounded by pigment cells and covered by translucent skin that lets in light, but without any lens or other refractive body. We may, however, descend even a step lower and find, according to the French biologist Monsieur S. Jourdain, clusters of pigment cells lacking nerves and resting merely in cytoplasm, apparently serving as crude organs for the perception of light and dark. In certain sea star species, small depressions in the layer of pigment surrounding the nerve are filled, as described by Monsieur Jourdain, with transparent gelatinous matter projecting with a convex surface like the cornea in higher animals. Monsieur Jourdain suggests that this serves not to form an image, but only to concentrate the luminous rays and facilitate their perception. In this concentration of the light rays we gain the first and by far the most important step towards the formation of a true, picture-forming eye; for if we place the naked extremity of the optic nerve—which in some of the lower animals lies deeply buried in the body and in others near the surface—at the right distance from the light-concentrating apparatus, an image will be formed on it.

Among the arthropods[5] and annelids we may start from an optic nerve simply coated with pigment, the latter sometimes forming a sort of pupil but without any lens or other optical contrivances. With insects we now know that

Figure 6.12 Compound eye of a fly. Note the hundreds of lenses at the eye's surface.

the numerous facets on the cornea of their great compound eyes form true lenses (Figure 6.12), and that their photosensitive receptors cells, the cones, include curiously modified nervous filaments. But these organs are so incredibly diversified among arthropods that the German biologist Fritz Müller formerly divided these animals into three main classes with seven subdivisions, in addition to establishing a fourth major class of arthropods with simple eyes.

When we reflect on the wide, diversified, and graduated range of structures that we find in the eyes of lower animals, and when we bear in mind how small the number of all living forms must be in comparison with all those that have become extinct, it becomes easier to believe that natural selection may well have gradually converted the simple apparatus of an optic nerve, coated with light-absorbing pigment and covered with a transparent membrane, into an optical instrument as perfect as is possessed by any members of the Arthropoda.

Anyone who has come this far with my argument ought not to hesitate in going one step further: if you find upon finishing this volume that large bodies of facts, otherwise inexplicable, can now be explained satisfactorily by the theory of gradual modification through natural selection, then you should admit that even a structure as perfect as an eagle's eye might thus be formed, even without knowing any of the transitional steps to the final product. Some have objected that in order to modify the eye and still preserve it as a perfect instrument, many changes would have to have happened simultaneously, which, it is assumed, could not be done through natural selection. But as I have attempted to show in my work on the variation of domestic animals, we need not suppose that the modifications were

---

[5] This group, members of the phylum Arthropoda, includes animals as diverse as insects, spiders, crabs, lobsters, and copepods.

all made simultaneously, as long as they were extremely slight and the process was gradual. Also, different kinds of modification can serve for the same general purpose. As my fellow naturalist Mr. Alfred Russel Wallace has remarked, "if a lens has too short or too long a focus, it may be amended either by an alteration of curvature, or an alteration of density; if the curvature be irregular, and the rays do not converge to a point, then any increased regularity of curvature will be an improvement. So the contraction of the iris and the muscular movements of the eye are neither of them essential to vision, but only improvements that might have been added and perfected at any stage of the construction of the instrument."

Even within the Vertebrata,[6] the highest division of the animal kingdom, we can start from an eye so simple that it consists—as in the lancelets (see Figure 4.16)—of a little sack of transparent skin, furnished with a nerve and lined with pigment, but destitute of any other apparatus. In fishes and reptiles, as Sir Richard Owen of the British Museum has remarked, "the range of gradations of dioptric (or refractive) structures is very great," while in humans, according to the high authority of the German doctor and biologist Rudolf Virchow, even our own beautiful crystalline lens is formed during embryonic development by an accumulation of epidermal cells lying in a sack-like fold of the skin, and the vitreous body is formed from embryonic subcutaneous tissue.

To arrive at a valid conclusion regarding eye formation, with all its marvelous yet not absolutely perfect characters, it is crucial that reason should conquer the imagination. But I have felt the difficulty far too keenly myself to be surprised at others hesitating to extend the principle of natural selection to so startling a length. Still, let us continue the argument.

It is scarcely possible to avoid comparing the eye with a telescope. We know that the telescope has been perfected by the long-continued efforts of the highest human intellects; naturally we infer that the eye has been formed by a somewhat analogous process. But might this inference be presumptuous? Have we any right to assume that the Creator works by intellectual powers like ours? If we must compare the eye to an optical instrument, we ought to imagine a thick layer of transparent tissue containing fluid-filled spaces, with a nerve sensitive to light beneath. Then suppose every part of this layer to be continually changing very slowly in density over time, so as to separate into layers of different densities and thicknesses, placed at different distances from each other, and with the surfaces of each layer slowly changing in form. Further we must suppose that there is a power, represented by natural selection or the survival of the fittest, always intently watching each slight alteration in the transparent layers, and carefully preserving any of these variants that tend to produce a more distinct image. We must suppose each new state of the instrument to be multiplied by the

---

[6] This subclass of the phylum Chordata includes all animals with vertebrae (e.g., birds, fish, reptiles, amphibians and mammals—including humans).

millions, each to be preserved until a better one is eventually produced. Then we must suppose the old ones to be all destroyed, probably through competition with individuals possessing the better ones.

In living organisms, natural variation will cause the slight alterations, reproduction through the generations will multiply them almost infinitely, and natural selection will then pick out each improvement with unerring skill. Let this process go on for millions of years, and during each year on millions of individuals of many kinds.[7] May we not then believe that a living optical instrument might eventually be formed that is as superior to one made of glass as are the works of the Creator to those of Man?

Modes of Transition

If someone could demonstrate the existence of even just one complex organ that could not possibly have been formed by numerous successive, slight modifications, my theory would absolutely break down. But I can find no such case. No doubt many organs exist of which we do not know the transitional forms, particularly if we look to much-isolated species around which, according to my theory, there has been much extinction. Or again, let's take an organ common to all the members of a class. In this case, the organ must have been originally formed very long ago, and all the many current members of the class must have been developed after that time. In order to discover the early transitional grades through which the organ in question has passed, we should have to look to very ancient ancestral forms, forms that have long been extinct.

We should be extremely cautious in concluding that an organ could not have been formed by transitional gradations of some kind. Many cases could be given among the lower animals of the same organ performing several wholly distinct functions at the same time. In dragonfly larvae and in the fish genus *Cobites*, for example, the digestive tract does three different things: it digests, it respires, and it excretes. And if the freshwater *Hydra* (Figure 6.13) is turned inside out, the inner and outer surfaces of the animal will exchange tasks: the former exterior surface will now digest, and the former stomach will now exchange gases. In such cases, natural selection might slowly lead to specialization, so that the whole organ or one part of an organ that had previously performed two or more functions might now perform one

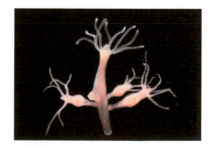

---

[7] Actually, the process may occur much more quickly than that: See Nilsson, D-E. and S. Pelger. 1994. A pessimistic estimate of the time required for an eye to evolve. *Proceedings of the Royal Society of London* Series B 256: 53–58.

Figure 6.13 This freshwater hydra is reproducing asexually by budding.

function alone, if any advantage would be gained from doing so. Thus, by many insensible steps, its nature would be greatly changed.

Consider a similar example with plants. Many plants regularly produce several differently constructed flowers; if such plants were to produce only one kind of flower, a great change in the character of the species would be brought about with comparative suddenness. It is, however, likely that the two sorts of flowers now borne by the same plant were originally differentiated by finely graduated steps, which may still be followed in a few cases.

We also know of instances in which two distinct organs, or the same organ in two very different forms, may simultaneously perform the same function in a single individual. This provides an extremely important means of future transition. Let me give one example. Some fish with gills take up the oxygen dissolved in water at the same time that they also take up oxygen in gaseous form from the air in their swim bladders (Figure 6.14), this latter organ being divided by highly vascularized partitions and being provided with air through a specialized pneumatic duct. In addition, the swim bladder of certain fishes has come to also function as an accessory to the organs of hearing. To give a similar example from the plant kingdom, plants climb by three distinct means—by spirally twining, by clasping a support with their sensitive tendrils, and by emitting aerial rootlets. These different mechanisms of climbing are usually found in distinct groups of plants; however, a few species exhibit two of the above mechanisms or even all three combined in the same individual. In all such cases, one of the two organs might readily be modified and perfected so as to perform all the work, being aided by the other organ during the progress of modification. This other organ—now freed of its responsibilities—might later become gradually modified for some other quite distinct purpose, or be wholly obliterated.

The illustration just given of the fish swim bladder is a good one, because it clearly shows us the highly important fact that an organ originally

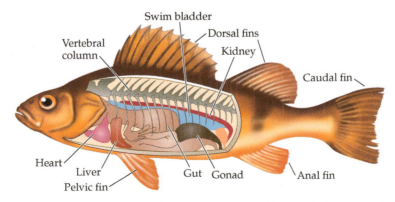

Figure 6.14 Fish swim bladder (shown in blue). Most fish use this organ to become neutrally buoyant in the water, after which they need not expend energy to stay at that particular depth.

constructed for one purpose (in this case for adjusting an animal's buoyancy), can in fact be converted into one for a widely different purpose (respiration). Indeed, all physiologists admit that the swim bladder is homologous[8] (that is, "ideally similar in position and structure") with the lungs of the higher vertebrate animals. Thus there is no reason to doubt that the swim bladder has actually been converted into lungs—an organ used exclusively for respiration.

According to this view, then, it may be inferred that all vertebrate animals with true lungs have descended through many generations of reproduction from an ancient and unknown prototype animal that was furnished with a buoyancy-control apparatus or swim bladder. We can thus understand the otherwise strange fact that every particle of food and drink that we swallow has to pass over the opening to our windpipe, and with some risk of falling into the lungs despite the beautiful contrivance by which our glottis (the opening between the vocal cords) is closed, as described by Sir Owen. And while it is true that in the higher vertebrates, including adult humans, the gills have wholly disappeared, yet we still see gill slits on the sides of the neck during embryonic development, along with the loop-like course of the arteries marking their former positions. Conceivably, the now utterly lost gills might instead have been gradually worked on by natural selection for some other distinct purpose. For instance, the German zoologist Hermann Landois has shown that the wings of insects develop from the tracheae;[9] thus it is highly probable that in this large and successful group of animals, organs that once served for respiration have actually been converted into organs for flight. What a magnificent thought!

In considering the evolutionary modification of organs, it is so important to bear in mind the probability of an organ being converted from one function to another that I will give one more example, this time from the world of barnacles. Stalked barnacles (Figure 6.15) have two minute folds of skin, which I have called "ovigerous frena." These folds produce a sticky secretion that holds the embryos in place within the parent's nursery chamber until they hatch from their egg membranes. These barnacles have no gills; gas

---

[8] Today that term is used more clearly to mean similarity by descent from a common ancestor. The contrasting term "analogous" refers to structures that are similar in structure or function, but without descent from a common ancestor.

[9] These are major components of the insect (and spider) gas exchange system, guiding air from small openings on the outside of the body to individual cells within the animal's tissues.

Figure 6.15 A stalked ("gooseneck") barnacle (*Pollicipes pollicipes*).

exchange is accomplished very simply, across the general body surfaces and those of the nursery chamber, together with the small frena (folds of membrane that restrict motion). On the other hand, the acorn barnacles, like members of the genus *Balanus*, for example (see Figure 5.8A) have no ovigerous frena; their embryos lie loose at the bottom of the nursery chamber, within the well-enclosed outer shell of the adult. They do have, however, in the same relative position as the frena of stalked barnacles, a set of large, much-folded membranes that freely communicate with the circulatory pouches of the animal's nursery chamber and outer body, and which have been considered by all naturalists to act as gills. Now I think that no one will dispute that the ovigerous frena in the one family of barnacles are strictly homologous with the gills of the other family; indeed, they graduate into each other. Therefore it need not be doubted that the two little folds of skin that had originally served as ovigerous frena, but which also very slightly aided in gas exchange, have been gradually converted by natural selection into gills, simply through an increase in their size and the obliteration of their adhesive glands. If all stalked barnacles had before now gone extinct—and they have indeed suffered far more extinction than have the sessile barnacles—then who would ever have imagined that the gills of the acorn barnacle had originally existed as organs whose role was to prevent the eggs from being washed out of the nursery chamber?

There is another possible mode of evolutionary transition to consider: namely, reproducing earlier or later in life. This idea has lately been promoted by the American paleontologist and comparative anatomist Professor Edward Drinker Cope and others in the United States. We now know that some animals can reproduce at a very early age, before they have acquired their perfect adult characters. If this power became thoroughly well-developed in a species, it seems probable that the presently existing morphologically distinct adult stage would sooner or later be lost from the life cycle. In that case, especially if the larva was very different-looking from the adult form, the character of the species would be greatly changed and simplified.[10]

Finally, we know that many animals, after arriving at maturity, go on changing in character during nearly their whole lives. With mammals, for instance, the form of the skull typically changes a great deal with age. The Scottish pathologist and naturalist Dr. James Murie has given some striking instances of this with seals. And of course everyone knows how the horns of stags become more and more branched with age, and that the plumes of some birds become more finely developed as each grows older. Professor Cope writes that the teeth of certain lizards also change much in shape with advancing years; and with crustaceans, some parts—both important ones and trivial ones—assume a new character after maturity, as recorded

---

[10] Indeed, this may well explain the evolution of planktonic "larvaceans" (solitary free-swimming organisms in the subphylum Urochordata, phylum Chordata) from immobile sea squirt ancestors.

by Fritz Müller. In such cases—and many more could be given—if the age for reproduction were delayed, the character of the species, at least in its adult state, would be modified. Nor is it improbable that the earlier stages of development would in some cases be hurried through and finally lost. Whether species have often—or even ever—been modified through this comparatively sudden mode of transition, I can form no opinion. But if this has in fact occurred, the differences between the young and the mature, and between the mature and the old, were probably acquired long ago by gradual steps.

Special Difficulties of the Theory of Natural Selection

Although we must be extremely cautious in concluding that any organ could not have been produced by successive, small, transitional gradations, several cases seem to point in that direction. But I believe that all of these cases can ultimately be explained through the power of natural selection and survival of the fittest acting over very long periods of time.

One of the most difficult situations to explain is that of neuter, nonreproductive insects, which often look very different from both the males and the fertile females of the same species. I will discuss that case in the next chapter. Here I will instead discuss another case of special difficulty for my theory: the electric organs of fishes.

It is impossible to conceive by what gradual steps these wondrous organs have been produced. But this is not surprising, for we do not even know with certainly of what use these organs are to their bearers. In the Amazonian electric knife fish (*Gymnotus* spp.) (Figure 6.16A), and the electric ray *Torpedo* (Figure 6.16B), the electric organs no doubt serve as a powerful means of defense, and perhaps also for securing prey; yet in the ray, as noted by the Italian neurophysiologist Carlo Matteucci, a similar organ in the tail manifests very little electricity, even when the animal is greatly irritated—so little, in fact, that it can hardly be of any use in either defense or prey capture.

(A)   (B)

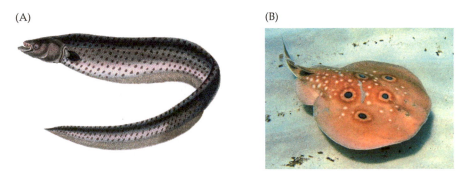

Figure 6.16 (A) The electric knife fish (*Gymnotus inaequilabiatus*).
(B) The electric ray (*Torpedo torpedo*).

Moreover, the Irish surgeon and anatomist Dr. Robert McDonnell has shown that besides the organ just referred to, there is another organ near the ray's head that is also not known to be electrical, but which appears nevertheless to be truly homologous with the electric battery in the electric ray. It is generally believed that there is a close similarity between these organs and ordinary muscles—in the fine structural details, in the distribution of the nerves, and in the manner in which they are acted on by various reagents. Remember, too, that muscular contraction is accompanied by electrical discharge. Moreover, as the English physician Dr. Charles Bland Radcliffe insists, "in the electrical apparatus of the torpedo during rest, there would seem to be a charge in every respect like that which is met with in muscle and nerve during rest, and the discharge of the torpedo, instead of being peculiar, may be only another form of the discharge which depends upon the action of muscle and motor nerve." Beyond this we presently know little. But as we know so little about the uses of these organs, and as we know nothing about the habits and structure of the ancestors of the existing electric fishes, it would be extremely bold to maintain that there is no mechanism by which these organs might have been gradually developed over time.

Electric organs appear at first to offer another and far more serious difficulty: they occur only in about a dozen kinds of fish, of which several are only distantly related to each other and to any of the others. When we find the same organ in several members of the same group, especially in members having very different habits of life, we may generally attribute its presence to inheritance from a common ancestor, and its absence in some of the members to loss through disuse or natural selection. If all electric organs had in fact been inherited from some one ancient ancestor, we might have expected that all existing electric fishes would be closely related to each other. But this is far from the case. Nor does the geological record suggest that most fishes formerly possessed electric organs, which most of their modified descendants have now lost.

So how can we account for this intriguing distribution of electric organs among fish? In looking at the subject more closely, we find that the electric organs of the different fish species are situated in different parts of the body. They also differ in construction among the different species, as well as in the arrangement of the disc-like plates that are stacked one on another like batteries, and, according to Italian anatomist Filippo Pacini, even in the mechanism through which the electricity is produced. Lastly, the electric organs of the different species are supplied with nerves proceeding from different sources: this is perhaps the most important of all the differences. Thus, the electric organs found in the various fish species cannot be considered as homologous, but only as analogous in function—they seem not to have arisen only once in a common ancestor; if they had, they would have closely resembled each other in all respects. Thus there is no need to explain how an identical organ arose separately in several remotely related species: the organs are *not* identical. But we still have a lesser yet still great difficulty to resolve: by what gradual steps have these organs been developed in each separate group of fishes?

The light organs that occur in a few insect species belonging to widely different families, and which are situated in different parts of the body in different species, offer, under our present state of ignorance, a difficulty almost exactly parallel to that of the electric organs of fishes. Other similar cases could be given. For instance, in plants we find what seems to be the same curious contrivance in both orchids and milkweeds (genus *Asclepias*), namely a mass of pollen grains borne on a footstalk with an adhesive gland, which sticks the pollen to the backs of insects for transport to other flowers. And yet these two groups are about as remotely related as is possible among flowering plants. But here again, the parts are not homologous (i.e., they do not come from a common ancestor).

In all cases of beings far removed from each other in the scale of organization but which are furnished with similar and peculiar organs, it will be found that although the general appearance and function of those organs may be the same, we can always detect fundamental differences between them. For instance, the eyes of cuttlefish, squid, and octopus (class Cephalopoda, in the phylum Mollusca) seem remarkably similar to those of vertebrate animals. In such distantly related groups no part of this resemblance can be due to inheritance from a common ancestor. The English biologist Mr. St. George Jackson Mivart has advanced this case as one of special difficulty, but I am unable to see the force of his argument. Any organ intended for vision must be formed of transparent tissue and must include some sort of lens for projecting an image at the back of a darkened internal chamber. Beyond this superficial resemblance, there is hardly any real similarity between the eyes of cephalopods and vertebrates, as may be seen by consulting Christian Victor Hensen's admirable memoir on these organs in cephalopods. I can't discuss this in detail here, but I may specify a few of the points of difference. The crystalline lens in the eyes of the most advanced cuttlefish (Figure 6.17) consists of two parts, placed one behind the other like two lenses, both having a very different structure and disposition to what we find among vertebrates. Moreover, the cuttlefish retina is wholly different, with a very clear inversion of the elemental parts, and with a large mass of nervous tissue included within the membranes of the eye. The relations of the muscles are also as different from those of the vertebrate eye as it is possible to conceive, and so on in many other points. It's really not

Figure 6.17 Cuttlefish belong to the same animal group that includes the octopus and the squid, and to the phylum that also includes the snails and oysters!

clear whether the same terms should even be used in describing the eyes of cephalopods and vertebrates. Anyone may, of course, deny that the eye in either case could have been developed through the natural selection of slight variations in many successive generations. But if one admits that this process produced the eye in one case, it is clearly possible as well in the other.

Fundamental structural differences in the visual organs of the two groups might have been anticipated, in accordance with this view of their manner of formation. Just as two people have sometimes independently hit on the same invention, so in the several foregoing cases it appears that natural selection, working for the good of each being and taking advantage of all favorable variations, has produced functionally similar organs independently in distinctly different groups, with none of the similarity owing to inheritance from a common ancestor.

Fritz Müller, in order to test the conclusions reached in this book, has carefully constructed a nearly similar line of argument. Several families of crustaceans—most of which are aquatic—include a few species that are physically modified to breathe air and live on land. In carefully studying the members of two of these families that are clearly closely related to each other, Mr. Müller found that the members of both species agree very closely in all important characters: in their sense organs and circulatory systems, in the position of the tufts of filtering hairs within their complex stomachs, and in the whole structure of the water-breathing gills—even to the microscopic hook by which they are cleansed. Thus, we might have expected that the equally important air-breathing apparatus would have looked the same in the land-dwelling species of both groups as well: for why should this one apparatus, which has an identical purpose in both groups of animals, have been made to differ, while all the other important organs are closely similar or even essentially identical?

Mr. Müller argues that this close similarity in so many points of structure must, in keeping with my views, be accounted for by inheritance from a common ancestor. But as the vast majority of the species in the above two families, as well as in most other crustaceans, are fully aquatic in their habits, it is improbable in the highest degree that their distant common ancestor should have been adapted for breathing air. Mr. Müller, thus being led to carefully examine the apparatus in the air-breathing species, found clear species-specific differences in several important points, including the position of the openings, the manner in which they are opened and closed, and in some other related details. Such differences are fully understandable, and might even have been expected, if we suppose that several species belonging to several distinct families had slowly and independently become adapted over many generations to live more and more out of water, and to breathe the air. For by belonging to different families, the different species would have differed to a certain extent; in keeping, then, with the principle that the nature of each variation depends on both the nature of the organism and that of the surrounding conditions, the different species would not have been expected to vary in exactly the same ways. Consequently, natural selection would have had different materials or variations to work on, so that even when eventually arriving at the same functional result, the structures thus acquired would almost necessarily have differed. In contrast, on the hypothesis of separate acts of creation, the whole case would remain unintelligible. This line of argument seems to have had great weight in leading Mr. Müller to accept the views that I have put forth in this book.

Figure 6.18 The parasitic varroa mite (*Varroa destructor*) on a honeybee host. Mites are arachnids, a group that also includes the spiders.

Another distinguished zoologist, the late Professor Jean Louis René Antoine Edouard Claparède of Switzerland, has argued in the same manner, and has arrived at the same result. He shows that there are parasitic mites (Figure 6.18) (members of the arachnid order Acari) that are all furnished with hair claspers, even though they belong to several distinct families and subfamilies. These organs must have been independently developed in the different species, as they differ too much from each other to have been inherited from a common ancestor: in some species they are formed by the modification of the forelegs; in others, through modifications of the hind legs; and in others, through modification of the maxillae or lips, or even of the appendages on the underside of the hind part of the body.

In all of these cases we see the same end gained and the same function performed in organisms that are not at all, or only remotely, related, and performed by organs that look similar but that develop quite differently in the different species. Indeed, it is a common rule throughout nature that even in closely related species, the same end is often achieved by very different means. The feathered wing of a bird is so differently constructed from the membrane-covered wing of a bat, and still more so are the differences between the four wings of a butterfly and the two wings of a fly, and the two wings with the forewings (called elytra) of a beetle. Bivalve shells are made to open and shut, but they do so through an amazing variety of patterns in hinge construction—from the long row of neatly interlocking teeth in the small marine clam *Nucula*, to the simple ligament of a blue mussel (Figure 6.19). Similarly, seeds are disseminated in such

Figure 6.19 Blue mussel (*Mytilus edulis*) showing how shells are joined along the edge by a simple strip of tissue called the ligament (pink).

a remarkable variety of ways: by their minuteness; by their capsule being converted into a light balloon-like envelope; by being embedded in pulp or flesh, formed of the most diverse parts and rendered nutritious, as well as conspicuously colored so as to attract and be devoured by birds; by having hooks of many kinds and even thin, serrated spines that adhere to the fur of passing dogs and other quadrupeds, or by being furnished with little wings and plumes, as different in shape as they are elegant in structure, so as to be wafted by every breeze.

I will give just one more example, as this subject of the same end being gained by the most diversified means well deserves our attention. Some authors maintain that the various animals and plants on this planet have been formed in many ways for the sake of mere variety, almost like toys in a toy shop, but such a view is not credible. For plants with separate sexes, the pollen clearly needs help in reaching another flower for fertilization; the same is also true for hermaphroditic plants (i.e., the flowers have both sexes) in which the pollen does not simply fall onto the stigma of that same flower (see Figure 4.7). In several species, the pollen grains themselves are light and incoherent, and may simply be blown by the wind through mere chance onto the stigma of another flower; indeed, this is the simplest plan one can imagine.

However, an almost equally simple, though very different plan occurs in many plants in which a symmetrical flower secretes a few drops of nectar, and is consequently visited by insects; the visiting insects then carry the pollen from the anthers to the stigma. From this simple stage of pollination we may pass through an inexhaustible number of contrivances, all functioning for the same purpose and effected in essentially the same manner, but entailing changes in every part of the flower. The nectar may be stored in variously shaped receptacles, for example, with the flower's stamens and pistils modified in many different ways, sometimes forming trip-like contrivances, and sometimes capable of neatly adapted movements through irritability or elasticity.

From such structures we may advance until we come to a case of truly extraordinary adaptation, as recently described by Dr. Hermann Crüger (Director of the Botanic Garden on the island of Trinidad) for the bucket orchids (genus *Coryanthes*) (Figure 6.20A). This orchid has part of its lower lip (the labellum, which is really a modified petal) hollowed out into a great bucket, into which drops of almost pure water continually fall from two secreting horns that stand above it; when the bucket is half full, the water overflows by a spout on one side. The base of the labellum lies over the bucket, and is itself hollowed out into a sort of chamber with two entrances at the sides; there are a number of curious fleshy ridges within this chamber. Now the most ingenious person, if he or she had not witnessed what takes place, could never have imagined what purpose all these parts serve. But Dr. Crüger saw crowds of large bumblebees visiting the gigantic flowers of this orchid. They visited not to suck nectar, but rather to gnaw off the ridges within the chamber above the bucket. While focused on their gnawing, they

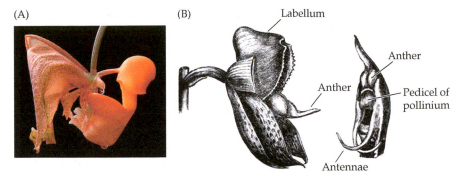

Figure 6.20 (A) A flower of the remarkable bucket orchid (*Coryanthes verrucolineata*). (B) Illustration of *Catasetum*.

frequently pushed each other into the bucket. With their wings being thus wetted they could not fly away, but were instead compelled to crawl out through the passage formed by the spout or overflow. Dr. Crüger saw a "continual process" of bees thus crawling out from their involuntary bath. Remarkably, the passage out is narrow, and is roofed over by the column, so that a bee, in forcing its way out, cannot help but rub its back against the sticky stigma and then against the sticky glands of the pollen masses. The pollen masses are thus glued to the back of whatever bee first happens to crawl out through the passage of a lately expanded flower, and are thus carried away by the bee. Dr. Crüger sent me a flower preserved in spirits of wine, along with a bee he had killed before it had quite crawled out; the dead bee still had a pollen mass fastened to its back. When the thus-encumbered bee flies to another flower, or to the same flower a second time, and is pushed by its comrades into the bucket and then crawls out through that narrow passage, the pollen mass must necessarily first contact the flower's sticky stigma and adhere to it…and the flower is thus fertilized.

Now at last we see the full use of every part of the flower: of the water-secreting horns, and of the bucket half full of water, which not only prevents the bees from flying away but also forces them to crawl out through the spout and rub against the precisely placed sticky pollen masses and sticky stigma. What an incredible system!

The construction of the flower in another closely related orchid in the genus *Catasetum* (Figure 6.20B) is widely different, although it serves the same end, and is equally curious. Bees visit these flowers, like those of *Coryanthe*, in order to gnaw the labellum. In doing so they inevitably touch a long, tapering, touch-sensitive projection that I have called the antenna. When touched, this "antenna" transmits a sensation or vibration to a certain membrane, which is instantly ruptured. This in turn sets free a spring by which the pollen mass is shot forth, like an arrow, in precisely the right direction, so that it adheres by its sticky extremity to the bee's back. The pollen mass of the male plant (for the sexes are separate in this orchid) is thus carried by the bee to the flower of the female plant, where it is brought into contact with the stigma. The stigma is sticky enough to break certain elastic threads: retaining the pollen thereby released, fertilization is thus accomplished.

How, it may be asked, in the foregoing and in innumerable other instances, can we understand the graduated scale of complexity and the multifarious means for gaining the same end in such a wide variety of organisms? **The answer no doubt is, as I have already remarked, that when two forms—which already differ from each other in some slight degree—continue to vary, the variability will not be of the same exact nature; the results obtained through natural selection for the same general purpose will consequently not be the same in both instances.** We should also bear in mind that every highly developed organism that we see has already passed through many changes, and that each modified structure tends to be inherited; thus each modification will not readily be lost, and may be again and again further altered. The structure of each part of each species, then, for whatever purpose it may serve, is the sum of many inherited changes through which the species has already passed during its successive adaptations to changed habits and changed conditions of life over many generations.

Even though in many cases it is most difficult even to conjecture about the transitions by which various organs have arrived at their present state, I have been astonished at how rarely we find any organ toward which no transitional grade is known to lead, especially considering how small the number of living species is compared with the number of those that have gone extinct in the past. It is certainly true that we never, or rarely, see new organs in any being appearing suddenly, as if they were created for some specific purpose; this reminds me of that well-noted old, but somewhat exaggerated canon of natural history, *Natura non facit saltum* ("Nature does not make jumps"). Indeed, we meet with this admission in the writings of almost every experienced naturalist. As the eminent French zoologist Henri Milne-Edwards has well expressed it, nature is prodigal in variety but stingy in innovation. **Why, based on the theory of Creation, would there be so much variety and yet so little real novelty?** Why should all the parts and organs of so many independent beings, each supposed to have been separately created for its proper place in nature, be so commonly linked together by graduated steps? Why should not nature take a sudden leap from structure to structure? On the theory of natural selection, we can clearly understand why nature should not do so, and in fact *cannot* do so: natural selection acts only by taking advantage of slight successive variations; she can never take a great and sudden leap, but must advance slowly by short and sure steps, over many, many generations.

Organs of Little Apparent Importance, as Affected by Natural Selection

As natural selection acts by life and death—by the survival of the fittest, and by the destruction of the less well-fitted individuals—I have sometimes felt great difficulty in understanding the origin or formation of parts of little importance, parts whose value seems too small to cause their preservation in successively varying individuals. It seems almost as great a problem, though of a very different kind, as in the case of the most perfect and complex organs such as eyes.

In the first place, however, we are much too ignorant with regard to the detailed functioning and ecological interactions of any one organic being to say what slight modifications might or might not be important to that organism. In Chapter 4 I gave instances of what seemed like very trifling characteristics, such as the down on the outside of fruit, the color of its flesh, and the color of the skin and hair of quadrupeds, which from being correlated with constitutional differences or from discouraging or encouraging the attacks of insects, might assuredly be acted on by natural selection. But the tail of the giraffe certainly looks like an artificially constructed flyflapper! At first it seems incredible that this tail could have been adapted for its present purpose by successive slight modifications, each better and better fitted for so trifling an object as to drive away flies! Yet we should pause before being too certain even in this case. In particular, we know that the distribution and existence of cattle and many other animals in South America absolutely depends on their ability to resist insect attacks; thus, individuals that could by any means defend themselves from these small enemies would be able to range into new pastures, thereby gaining a great advantage over others. It is not that the larger quadrupeds are actually destroyed (except in some rare cases) by flies, but that they are incessantly harassed and their strength reduced, so that they are more vulnerable to disease, or less able to search for food during a future famine, or to escape from predators.

Moreover, organs now of trifling importance may well have been of great importance to an early ancestor, and, after having been slowly perfected in some former period, were then transmitted through the generations to existing species in nearly the same state, even though they may now be of little use; any injurious changes to their structure would of course have been checked by natural selection. We may, for example, perhaps account for the general presence of tails and their use for so many purposes in many land animals—which in the structure of their lungs or modified swim bladders betray their aquatic origins—by noting how important the tail is as an organ of locomotion in most aquatic animals, Once a well-developed tail has been formed in some aquatic animal, it might subsequently come to be modified for all sorts of purposes...as a flyflapper, for example, or as an organ for grasping, or as an aid in turning, as in the case of dogs, although the degree to which the tail helps a dog turn must be slight, for a rabbit can double about even more quickly than a dog with hardly any tail at all.

On the other hand, we may easily be mistaken in attributing importance to certain characteristics and in believing that they have been developed through natural selection. We must by no means overlook the effects of so-called spontaneous variations, which seem to depend little on the nature of the surrounding conditions, or the tendency of organisms to sometimes revert to long-lost characteristics, or the effects of the complex "laws of growth," such as correlation, compensation, and the effects of pressure of one part on another,[11] as discussed in Chapter 5. Nor should we overlook the role of sexual selection, through which characteristics of use to one sex are often gained and then transmitted more or less perfectly to the other sex, even though they are of no use to that other sex. But structures thus indirectly gained, although at first of no advantage to a species, may subsequently have been taken advantage of by its modified descendants, under new conditions of life and with newly acquired habits.

For example, if all the woodpeckers we knew of were green, and we did not know that there were also many black or black and white kinds (often with a patch or two of red), I dare say that we should have thought that the green color was a beautiful adaptation to conceal these tree-frequenting birds from their enemies, and consequently that it was a characteristic of considerable importance, one that had been acquired through natural selection; but in fact, the color probably results mostly from sexual selection. Similarly, a trailing palm plant in the Malay Archipelago climbs the loftiest trees by using exquisitely constructed hook clusters around the ends of its branches. No doubt this contrivance is extremely useful to the plant, but the fact remains that we see very similar hooks on many plants that are not climbers, and which seem to serve instead as a defense against browsing quadrupeds. Thus, the spikes on the palm plant may at first have been developed for this defensive function, and subsequently have been improved and taken advantage of by the plant over many thousands of generations as it underwent further modification and became a climber.

As a third example, the naked skin on the head of a vulture (Figure 6.21A) is generally considered as a direct adaptation for wallowing in putridity. Or perhaps it is due to the direct action of the decomposing and putrifying matter on which it feeds. But we should be very cautious in

(A)

(B)

Figure 6.21 (A) The head of a turkey vulture (*Cathartes aura*). (B) The turkey (*Meleagris gallopavo*).

---

[11] Once again, we see that Darwin knows nothing about the genetic basis of variation—poor fellow!

drawing any such inference; after all, the skin on the head of the clean-feeding male turkey is also naked (Figure 6.21B), even though these birds feed mostly on grains, seeds, nuts, leaves, and small insects.

Lastly, the sutures in the skulls of young mammals have been proposed to be beautiful adaptations for easing the movement of babies through the birth canal at birth. No doubt they do facilitate that process, and may even be indispensable for it, but sutures also occur in the skulls of young birds and reptiles, which have only to escape from a broken egg; thus we may infer that this structure has arisen originally only from the natural laws of growth and has simply then been taken advantage of by mammals for childbirth.

We are profoundly ignorant of the cause of each slight variation or individual difference. We are immediately made aware of this by reflecting on the differences between the breeds of domesticated animals found in different countries, particularly in the less civilized countries where there has been but little methodical selection for any particular traits. Animals kept by savages in different countries often have to struggle to find food and are thus to some extent exposed to the forces of natural selection; individuals with slightly different constitutions would succeed best under different climates. With cattle, susceptibility to the attacks of flies is correlated with body color, as is the likelihood of being poisoned by certain plants; thus even color would be subjected to the action of natural selection. Some observers are convinced that a damp climate affects the growth of the hair, and that with the hair the animal's horns are correlated. Mountain breeds always differ from lowland breeds; a mountainous country probably affects the hind limbs by exercising them more, and possibly even affects the form of the animal's pelvis. If so, then by the law of homologous variation, the front limbs and the head would probably also be affected. The shape of the pelvis might also affect, by pressure, the shape of certain parts of the young in the womb. The laborious breathing necessary in high-altitude regions tends to increase the size of the chest; again, the correlation with other body parts would come into play. The effect of lessened exercise together with abundant food on the whole organization is probably still more important, and this in fact, as the German animal breeder Hermann von Nathusius has lately shown in his excellent treatise, is apparently one chief cause of the great modification that the various pig breeds have undergone.

But we are far too ignorant to speculate on the relative importance of the several known and unknown causes of variation. I have made the preceding remarks only to show that if we are unable to account for the characteristic differences in our several domestic breeds, which nevertheless are generally admitted to have arisen through ordinary reproductive processes from one or at most a few parent stocks, we ought not to lay too much stress on our ignorance of the precise cause of the slight analogous differences between true species.

Utilitarian Doctrine, How Far True? Beauty, How to Explain It?

The previous remarks lead me to say a few words about the protest lately made by some naturalists against my idea that every detail of structure has been produced for the good of its possessor. They believe instead that many structures have been created for the sake of beauty, to delight man or the Creator—a point beyond the scope of scientific discussion—or for the sake of mere variety, a view that I have already discussed. Such doctrines, if true, would be absolutely fatal to my theory.

I fully admit that many structures are now of no direct use to their possessors, and may never have been of any use to the ancestors of those individuals, but this does not prove that they were formed solely for beauty or variety. No doubt the definite action of changed conditions, and the various causes of modifications, which I have already discussed, have all produced an effect—probably a great effect—independently of any advantage thus gained. **But a still more important consideration is that most of the organization of every living creature is due to inheritance.** Consequently, though each being is assuredly well fitted for its place in nature, many structures now have no very close and direct relation to present habits of life. Thus, we can hardly believe that the webbed feet of the upland goose of South American grasslands, or of the frigate bird, which cannot swim or even walk well, and which takes most of its food in flight, are of special use to these birds. And it seems so unlikely that the similar bones in the arm of the monkey, in the foreleg of the horse, in the wing of the bat, and in the flipper of the seal, are of special use to these animals; we may safely attribute these structures to inheritance. But webbed feet no doubt were as useful to the distant ancestor of the upland goose and of the frigate bird as they now are to the most aquatic of living birds.

Thus it is logical to believe that the ancient ancestor of the seal did not possess a flipper, but rather a foot with five toes fitted for walking or grasping, and we may further venture to believe that the several bones in the limbs of the monkey, horse, and bat were originally developed on the principle of usefulness, probably through the reduction of more numerous bones in the fin of some ancient fish-like ancestor of the whole class. It is scarcely possible to decide how much allowance ought to be made for such causes of change as the definite action of external conditions, so-called spontaneous variations, and the complex laws of growth;[12] but with these important exceptions, we may conclude that the structure of every living creature either now is, or was formerly, of some direct or indirect use to its owner.

With respect to the belief that organisms have been created beautiful for the delight of people, let me first note that the sense of beauty obviously depends on the nature of the mind, irrespective of any real quality

---

[12] Again, Darwin would have so much enjoyed, and profited from, an introductory course in genetics. Imagine the look on his face if he sat in on such a class and first learned about genes.

(A)   (B)

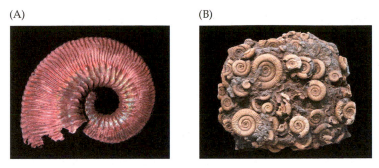

Figure 6.22 (A) A pyritized fossil ammonite (*Kosmoceras* sp.) from the Jurassic period. (B) A group of fossilized ammonites.

in the admired object: the idea of what is beautiful is not innate, and is not unalterable. We see this, for instance, in the men of different races admiring an entirely different standard of beauty in their women. If beautiful objects had been created solely for our gratification, it ought to be shown that there was less beauty on the face of the earth before we appeared than since we came on the stage. Were the beautiful snail shells, such as the volutes and cone snails, from tens of millions of years ago in the Eocene epoch, and the gracefully sculpted ammonites (Figure 6.22), which went extinct some 65 million years ago, all created so that we might—ages afterwards—admire them in our curio cabinets? Similarly, few objects are more beautiful than the microscopic silica cases of diatoms (Figure 6.23); were these created so that they might someday be examined and admired by us under the higher powers of microscopes?

And what about flowers? Flowers must be among the most beautiful productions of nature; but they have been made conspicuous in contrast to the green leaves around them, and consequently also beautiful, only so that they may be easily observed by the insects that are needed to complete their life cycles. I have come to this conclusion from finding it an invariable rule that flowers that are routinely fertilized by the wind are never gaily colored. Moreover, several plants habitually produce two kinds of flowers: one that is open and colored, to attract insects, and the other closed, not colored, destitute of nectar, and never visited by insects. Thus we

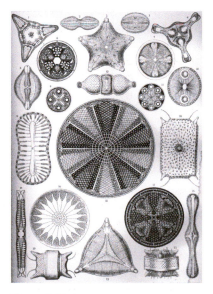

Figure 6.23 An assortment of diatoms.

Figure 6.24 Holly berries.

may conclude that if insects had never appeared on the face of the earth, our plants would not now be decked with beautiful flowers, but would have produced only such poor flowers as we see on fir, oak, nut, and ash trees, and on grasses, spinach, docks (*Rumex obtusifolius*), and nettles, all of which are fertilized by the wind.

A similar line of argument holds for fruits. Everyone can agree that a ripe strawberry or cherry is as pleasing to the eye as to the palate, and that the gaily colored fruit of the spindle-wood tree and the scarlet berries of the holly (Figure 6.24) are beautiful objects. But this beauty serves merely as a guide for birds and beasts, in order that the fruit may be eaten and the matured seeds then disseminated by the eater. Indeed, I infer that this is the case from having as yet found no exception to the rule that seeds embedded within a fruit of any kind (that is, within a fleshy or pulpy envelope) are always disseminated in this way as long as the fruit is colored in any brilliant tint, or rendered conspicuous by being white or black.

On the other hand, I willingly admit that a great number of male animals—as with all our most gorgeous birds, some fishes, reptiles, and mammals, and a host of magnificently colored butterflies—have been rendered beautiful for beauty's sake. But this has resulted through sexual selection, not for our delight; that is, by the more beautiful males having been continually preferred by the females. So it is with the music of birds. We may infer from all this that a nearly similar taste for beautiful colors and musical sounds runs through a large part of the animal kingdom. When the female is as beautifully colored as the male, which is often the case with birds and butterflies, the cause apparently lies in the colors acquired through sexual selection having been transmitted to both sexes, instead of to the males alone. How the sense of beauty in its simplest form—that is, the reception of a peculiar kind of pleasure from perceiving certain colors, forms, and sounds—was first developed in the mind of humans and of the lower animals, is a very obscure subject. The same sort of difficulty is presented if we ask how it is that certain flavors and odors give us pleasure, while others displease us. Habit appears to play a role in all these cases, but there must be some fundamental cause in the constitution of the nervous system of each species.

Natural selection cannot possibly produce any modification in a species exclusively for the good of another species, even though throughout nature one species incessantly takes advantage of, and profits by, the structures of others. But natural selection can, and does, often produce structures for the direct injury of other animals, as we see, for example, in the fang of the adder, and in the ovipositor of the ichneumon wasps (Figure 6.25) (insect family Ichneumonidae), which it uses to deposit its eggs into the living bodies of other insects.

If it could be proven that any part of the structure of any one species had been formed for the exclusive good of another species, it would annihilate my theory, for such could not have been produced through natural selection. Although many works on natural history claim that some structures in one species do indeed serve for the exclusive benefit of a different spe-

Figure 6.25 Ichneumon wasp (*Xorides praecatorius*) with prominent ovipositor (bottom).

cies, I cannot find even one example that seems to me to hold any weight. Most authors admit that the rattlesnake has a poison fang for its own defense, and for killing its prey. But some of these same authors also believe that the rattlesnake is furnished with a rattle to deliberately harm itself by warning its prey. I would almost as soon believe that the cat curls the end of its tail when preparing to spring in order to warn the doomed mouse of its coming fate!

It is much more likely that the rattlesnake uses its rattle, the cobra expands it frill, and the puff adder swells while hissing loudly and harshly in order to alarm the many birds and beasts that are known to attack even the most venomous species. Snakes act on the same principle that makes the hen ruffle her feathers and expand her wings when a dog approaches her chickens…but I do not have space here to say more about the ways that animals endeavor to frighten away their enemies.

Natural selection will never produce—in any organism—any structure that injures that organism more than it provides a benefit to that organism, for natural selection acts solely by and for the good of each individual. No organ will be formed, as Reverend William Paley has remarked, for the purpose of causing pain or for injuring its owner. If a fair balance be struck between the good and evil caused by each part, each part will be found on the whole to be advantageous to its possessor. Over long periods of time, under changing conditions of life, if any part comes to be injurious it will be modified; if not, the organism will become extinct, as indeed myriads have become extinct in the past.

Natural selection tends only to make each organism as perfect as, or slightly more perfect than, the other inhabitants of the area in which it competes. And we see that this is the standard of perfection attained under nature. The endemic native plants and animals of New Zealand, for instance, are perfect when compared with each other; but their populations are now rapidly dwindling before the advancing legions of plants and animals that have been introduced from Europe.

No, natural selection will not produce absolute perfection, and we do not often see, as far as we can judge, such a high standard met under nature. According to Mr. Müller, for example, the correction for the aberration of light is not perfect even in that most perfect organ of sight, the human eye. The German physician and physicist, Hermann von Helmholtz, whose judgment no one will dispute, describes in the strongest terms the wonderful powers of the human eye, but then adds these remarkable words: "That which we have discovered in the way of inexactness and imperfection in the optical machine and in the image on the retina, is as nothing in comparison with the incongruities which we have just come across in the domain of the sensations. One might say that nature has taken delight in accumulating contradictions in order to remove all foundation from the theory of a pre-existing harmony between the external and internal worlds." If our reason leads us to admire with enthusiasm a multitude of inimitable contrivances in nature, this same reason tells us, though we may easily err on both sides, that some other contrivances are less perfect. Can we consider the sting of the bee to be perfect, considering that when used against many kinds of enemies it cannot be withdrawn, because of the backward-pointing teeth along its length, and so causes the death of the bee by tearing out its entire digestive tract?

But if we look at the bee's stinger (Figure 6.26) as having existed in a remote ancestor as a boring and serrated instrument,[13] like that seen in so many other hymenopterans, and that it has since been modified—but not perfected—for its present purpose, with the poison originally adapted for some other purpose, such as to produce galls, and since intensified, we can perhaps understand how it is that the use of the sting should so often cause the insect's death: for if the power of stinging is generally useful to the social community, it will fulfill all the requirements of natural selection even though it may cause the death of some few members.

And again, if we admire the truly wonderful power of scent by which the males of many insects find their females for mating, can we also admire the production for this single purpose of thousands of male drones, which are utterly useless to the community for anything other than to fertilize the eggs of a receptive queen, and which are ultimately slaughtered by their industrious and sterile sisters? This is certainly not evidence of perfection in nature. It may be difficult, but we ought also to admire the savage instinctive hatred of the queen bee, which urges her to destroy the young queens—her own daughters—as soon as they are born, or to perish herself in the combat, for undoubtedly this is good for the entire community; maternal love or maternal hatred, though the latter fortunately is most rare, is all the same to the inexorable principle of natural selection. And if we admire the several ingenious contrivances by which orchids and many other plants are

---

[13] Actually, the stinging structure was probably used for egg laying originally, which would explain why only female bees can sting.

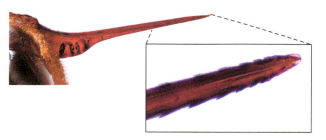

Figure 6.26 A bee stinger removed from a honeybee, lateral view. Inset, barbs are more apparent when viewed from above.

fertilized by insects, can we consider as equally perfect the release of dense clouds of pollen by our fir trees, so that just a few granules may be wafted by chance onto the ovules of another tree?

Summary

In this chapter I have discussed some of the difficulties and objections that may be brought against my theory of natural selection. Although many of them are serious, I think that in the preceding discussion I have thrown light on several facts that a belief in numerous independent acts of special creation are utterly unable to explain. We have seen that species at any one period are not indefinitely variable and are not linked together by a multitude of intermediate gradations, partly because the process of natural selection is always very slow and acts only on a few forms at a time, and partly because the very process of natural selection implies the continual supplanting and extinction of the preceding and intermediate steps. Closely allied species, now living throughout a large, continuous area, must often have been formed when the area was broken up into isolated sections, and when the conditions of life did not sensibly graduate away from one part to another. When two varieties of a species are formed in different parts of a continuous area, an intermediate variety will often be formed as well, fitted for life in the intermediate zone between the two varieties; but as I have explained, the intermediate variety will usually exist in smaller numbers than the two forms it connects. In consequence, during the course of further modification, the two major forms, because they exist in larger numbers, will have a great advantage over the less numerous intermediate variety, and will thus generally succeed in eventually supplanting and exterminating it.

We have seen in this chapter how cautious we should be in concluding that the most different habits of life could not graduate from one to the other—that a bat, for instance, could not have been formed by natural selection from an animal that at first only glided through the air.

We have also seen that a species under new conditions of life may change its habits. Or it may have a variety of habits with some very unlike those of its closest relatives in the same family. We can understand, then, bearing in mind that each living being is *trying* to live wherever it *can* live, how it has come about that there are fully terrestrial upland geese with webbed feet, and ground-living woodpeckers, and diving thrushes, and petrels with the habits of open ocean auks.

The belief that an organ as perfect as the human eye could ever have been formed gradually by natural selection is enough to stagger anyone. And yet, in the case of the vertebrate eye or any other organ, if we know of a long series of gradations in complexity, each good for its possessor, then, under changing conditions of life there is no logical obstacle to eventually acquiring any conceivable degree of perfection through natural selection. Even in those cases in which we don't know of any intermediate or transitional states, we should be extremely cautious in concluding that none existed; the transformation of many organs shows what wonderful changes in function are at least possible. The fish's swim bladder, for example, has apparently been converted into an air-breathing lung. In some ancient variety, the same organ must have simultaneously performed two very different functions (buoyancy control and gas exchange) in the same individual, and then gradually become specialized for just one of those functions—gas exchange. In many other cases, two distinct organs simultaneously performed the same function, opening the path, through redundancy, for one of the organs to then transition to a very different function.

We have seen that in two distantly related organisms, organs having the same function and closely resembling each other in external appearance may have been separately and independently formed; if so, essential differences in their structure can almost always be detected when such organs are examined closely. This naturally follows from the principle of natural selection. The common rule throughout nature is "Infinite diversity of structure for gaining the same end," which again follows from the same great principle of natural selection.

In many cases we are far too ignorant to claim that any particular part or organ is so unimportant for the welfare of a species that modifications in its structure could not have been slowly accumulated through natural selection. In many other cases, modifications probably result directly from the laws of variation or of growth, independently of any benefits having been thus gained. But even such structures have probably often been further modified by natural selection for the good of individuals under new conditions of life. It seems also likely that a part formerly of high importance to an organism has frequently been retained (such as the tail of an aquatic animal by its now-terrestrial descendants), even though it has now become of such small importance that it could not, for its present function, have been acquired by means of natural selection.

Natural selection can produce nothing in one species for the exclusive good or injury of another, although it may well produce parts, organs, and excretions (such as the sugar-containing honeydew of aphids) highly useful or even indispensable to other species, or even injurious to another species, as long as those parts, organs, or secretions are at the same time useful to the possessor. Wherever population sizes are large, natural selection acts through competition between the inhabitants and consequently leads to successes and failures in the battle for life in that particular region or area. Thus the inhabitants of one region, generally the smaller one, often yield to the inhabitants of another region (generally the larger one), and for this simple reason: the larger region will be home to more individuals and more diversified forms, and the competition will thus have been more severe there, so that the standard of perfection will have been rendered higher. Natural selection will not necessarily lead to absolute perfection, nor, as far as we can judge by our limited faculties, can absolute perfection be everywhere expected.

On the theory of natural selection we can clearly understand the full meaning of that old rule in natural history, *Natura non facit saltum*—"Nature does not make jumps." This rule, if we look only at the present inhabitants of our world, is not strictly correct. But if we include all those ancestors from past times, whether known or unknown, the rule must, on the theory of natural selection, in fact be strictly true. Organisms evolve gradually over long periods of time.

The Law of Unity of Type and of the Conditions of Existence Embraced by the Theory of Natural Selection

It is generally believed that all living beings have been formed on two great laws: the Unity of Type, and the Conditions of Existence. Unity of Type, as promoted by Étienne Geoffroy Saint-Hilaire and Sir Richard Owen, refers to the fundamental similarities in structure that we see in organisms found within the same taxonomic group, regardless of their lifestyles. My theory of natural selection explains unity of type quite simply, by descent from a common ancestor.

The second law, Conditions of Existence, as promoted so often by the French anatomist Georges Cuvier, refers to the way in which many structures—the hands of people, the wings of birds, the flippers of a seal, for example—are so marvelously adapted for the particular lifestyles of their owners, as though their lifestyle (i.e., their "conditions of existence") directly controlled their anatomy. This law is again fully explained by natural selection: natural selection acts by either now slowly adapting the varying parts of each being to its lifestyle and passing those adaptations along to offspring, or by the ancestors having adapted those parts to that lifestyle in the past. Thus, in fact, the Conditions of Existence law is the higher law, as it includes, through the inheritance of former variations and adaptations that were selected for in ancestors, that of Unity of Type.

Key Issues to Talk and Write About

1. Find out two interesting things about one of the people that Darwin mentions in this chapter. Choose from the following:

Alphonse de Candolle	Edward Drinker Cope
Edward Forbes	Fritz Müller
Hewett Cottrell Watson	Carlo Matteucci
Asa Gray	Filippo Pacini
Thomas Vernon Wollaston	Henri Milne-Edwards
Samuel Hearne	St. George Jackson Mivart
Félix de Azara	Rudolf Virchow
John Avebury Lubbock	Hermann von Nathusius
Hermann von Helmholtz	Georges Cuvier
Henri Louis Frédéric de Saussure	Jean Louis René Antoine Edouard Claparède

2. Picture a cow: slow moving, feeding on grass, and occasionally mooing. Based on your reading of this chapter, try to come up with a scenario in which a cow-like animal in the wild might eventually, over many thousands of generations, become adapted for feeding on frogs.

3. According to Darwin, why do we rarely see intermediate stages between an ancestral form and a modern form adapted for a very different lifestyle (e.g., winged birds and their presumably non-winged ancestors?)

4. Darwin argues that although we can't directly trace the exact steps through which a complex organ like the eye progressed from some ancient ancestor of today's eyed species, we can get a good sense of what those steps may have been by looking at how that organ is developed today in different species. How convincing do you find this argument? Explain your reasoning.

5. How might redundant parts or organs have played a role in the gradual evolution of novel structures and novel functions, according to Darwin?

6. According to Darwin, how should we determine whether similar organs in very different species were independently evolved, or passed down and evolved from a common ancestor? Give one example of his argument.

7. Using Darwin's comparison of the cephalopod and vertebrate eyes, tell a classmate or friend how natural selection can explain the appearance of similarly specialized organs in two completely different groups of animals.

8. Briefly summarize Darwin's argument about why we should not expect natural selection to bring about perfection in any of the traits that it shapes.

9. What are the male parts of a flower called? What are the female parts of a flower called? Which of Darwin's descriptions of fertilization among orchids do you find most interesting, and why?

10. List all of the animals and plants that Darwin uses as examples in this chapter.

11. Write a stand-alone, accurate, and informative sentence that summarizes the essential content of one of the following paragraphs:

    a. "We also know of instances in which two distinct organs..." (see page 158)

    b. "For example, if all the woodpeckers we knew of were green..." (see page 170)

    c. "And what about flowers? Flowers must be among the most beautiful productions of nature..." (see page 173)

12. Try rewriting these sentences, to make them clearer and more concise:

    a. In the foregoing cases, we see the same end gained and the same function performed, in beings not at all or only remotely allied, by organs in appearance, though not in development, closely similar.

    b. But in fact, the color is probably mostly the result of sexual selection.

    c. The pollen does not spontaneously fall on the stigma of the flower.

Online Resources *available at* **sites.sinauer.com/readabledarwin**

Videos

6.1 Flying fish

6.2 Flying lemurs in Borneo

6.3 Flying lemurs in the Philippines

6.4 American mink

6.5 Electric knife fish

6.6 Torpedo ray fish

6.7 Bucket orchid pollination

6.8 *Catasetum fimbriatum* pollination mechanism

6.9 An orchid explosion

6.10 Frigate birds

6.11 Puff adder attacks mouse

Suggested Readings

See the following papers for fascinating information about the remarkable eyes of brittle stars:

Aizenberg, J., A. Tkachenko, S. Weiner, L. Addadi, and G. Hendler. 2001. Calcitic microlenses as part of the photoreceptor system in brittlestars. *Nature* 412: 819–822.

Aizenberg, J. and G. Hendler. 2004. Designing efficient microlens arrays: Lessons from Nature. *Journal of Materials Chemistry* 14: 2066–2072.

7

Miscellaneous Objections to the Theory of Natural Selection

In this chapter Darwin addresses a number of major criticisms that had been thrown at his theory of evolution by natural selection and survival of the fittest, particularly those delivered by zoologist St. George Jackson Mivart, who had initially been a strong supporter of the theory. In doing so, Darwin gets to talk in detail about some remarkable examples that he hasn't mentioned before, including the evolution of baleen whales from toothed ancestors, the evolution of climbing in plants, and the evolution of breasts in mammals. This chapter appeared in The Origin *for the first time in the Sixth Edition, on which this book is based.*

In this chapter I will consider various miscellaneous objections against my views, to make some of the previous discussions a little clearer. But it would not be useful to discuss all of the objections, as many have been made by writers who have not taken the trouble to understand the subject.

For example, a distinguished German naturalist has asserted that the weakest part of my theory is that I consider all organisms to be imperfect. However, what I really said is that all are not as perfect as they might have been in relationship to the sorts of lives they lead. This is clearly shown to be the case just from the number of native forms in many parts of the world that have yielded their places to intruding organisms—invasive or introduced species—from other countries; if the native species had been perfectly adapted to their surroundings, then the invading species wouldn't have been able to get a foothold. Nor can any organisms—even if they had been perfectly adapted to their niches at one time—have remained so perfectly adapted when environmental conditions changed unless they themselves also changed; and no one will dispute that the physical conditions of each country, as well as the numbers and kinds of its inhabitants, have undergone many such changes.

Another critic has recently insisted, with some parade of mathematical accuracy, that since longevity must be a great advantage to all species, anyone who believes in natural selection "must arrange his genealogical tree" so that all descendants have longer lives than any of their ancestors did! Can our critic not conceive that a biennial plant or one of the lower animals might range into a cold climate and perish there every winter, and yet, owing to the advantages gained through natural selection, survive as a species from year to year by means of its seeds, or its eggs? The English biologist Mr. Edwin Ray Lankester has recently discussed this subject, concluding that longevity probably varies with the relative structural and physiological complexity of each species, as well as with the amount of energy expended in its reproduction and general activity. These conditions have probably been determined largely through natural selection.

Some have also argued that, as none of the animals and plants of Egypt that we know about have changed during the last 3,000 or 4,000 years, probably none have changed in any other part of the world either. But as the English philosopher Mr. George Henry Lewes has remarked, although the ancient domestic races illustrated on various Egyptian monuments, or embalmed, are indeed closely similar or even identical with those now living, all naturalists nevertheless admit that such races have been produced through the prior modification of their original types. Moreover, Egypt is not typical of other parts of the world: in Egypt, the conditions of life seem to have remained absolutely uniform during the last several thousand years. The many animals that have remained unchanged since the end of the most recent glacial period would have been an incomparably stronger case to offer, for those organisms have been exposed to great changes of climate and have migrated over great distances. The fact of little or no modification having taken place in those organisms since the glacial period would have been of some use against those who believe in an innate and continuous law of development, but has little to do with the doctrine of natural selection or survival of the fittest. **Natural selection doesn't imply continuous change, only that when appropriate variations or individual differences of a beneficial nature happen to arise within a population, those will be preserved under certain favorable circumstances.**

The celebrated paleontologist Heinrich Georg Bronn, at the close of his translation of *The Origin* into German, asks, "How can the principle of natural selection allow a variety to lie side by side with the parent species?" I can answer this easily: If both have become fitted for slightly different habits of life or conditions, they might live together perfectly well, since they will not be directly competing for the same resources. And if we put "polymorphic species" on one side (in which several distinctly different body types are found within a single population), and all mere temporary variations on the other (such as differences in size, degree of albinism and so forth), the more permanent varieties are generally found inhabiting very distinct habitats—such as high land versus low land, or dry versus moist districts. Moreover, in the case of animals that wander about a good deal and cross freely when mating, their varieties seem to be generally confined to distinct regions, so that interactions between them are minimal.

Modifications not Necessarily Simultaneous

Mr. Bronn also insists that distinct species never differ from each other in just single characteristics, but rather in many parts. He then asks, how is it possible that so many parts of the organization should have been modified simultaneously through variation and natural selection? But why must we assume that all parts of an organism have been modified simultaneously? As I have noted earlier, the most striking modifications, excellently adapted for some particular function, might well be acquired by successive slight variations, first in one part and then in another. And as those variations would be transmitted all together to the next generation, it would look to us as though they had been simultaneously developed, although that was not actually the case. The best answer to the above objection, however, is afforded by those domestic races that people have modified for some special purpose, chiefly through our powers of selection over time. Look at the racehorse and the cart horse (see Figure 4.10) for example, or at the greyhound and the mastiff: over many generations we have modified their whole frames, and even their mental characteristics, but if we could trace each step in the history of their transformation—and the latter steps can indeed be so traced—we would not see great and simultaneous changes, but first just one part and then another slightly modified and improved. Even when we have applied artificial selection to some one characteristic—of which our cultivated plants offer the best examples—we invariably find that although this one part—whether it be the flower, the fruit, or the leaves—has been greatly changed over many generations, almost all the other parts have been slightly modified as well. These modifications have presumably taken place through the principle of correlated growth, along with so-called spontaneous variation—mechanisms that as I have explained earlier we understand very little about.

Modifications that are Apparently of No Direct Service

A much more serious objection has been put forth by Mr. Bronn, and even more recently by the French anatomist Pierre Paul Broca: Many characteristics appear to be of no use whatever to their owners, and therefore could not have been selected for. Mr. Bronn gives as examples the length of the ears and tails in the different species of hares and mice, along with the complex folds of the enamel in the teeth of many animals and a multitude of analogous cases. With respect to plants, the subject has been admirably discussed by the Swiss botanist Carl Wilhelm von Nägeli. Although Mr. von Nägeli admits that natural selection has accomplished much, he insists that the families of plants differ from each other mainly in morphological characteristics that seem to be quite unimportant for the welfare of the species. He consequently believes that plants show an innate tendency toward an automatic, progressive, and more perfect development over time. As examples of cases in which natural selection could not have acted, he offers the arrangement of cells in the plant's tissues, and of the leaves on the axis. To these we might add such items as the numerical divisions in the parts of the flower, the position of the ovules (the structure that produces and contains the female reproductive cells), and the shape of the seeds when they seem of no use for dissemination.

There is much force in the above objections. Even so, we ought to be extremely cautious in pretending to know what structures are now, or have been, of use to any species. There must be some efficient cause for each slight individual difference, as well as for the more prominent variations that occasionally arise; and if the unknown cause were to act persistently, it is almost certain that all the individuals of the species would be similarly modified. With respect to the assumed uselessness of various body parts and organs, it is hardly necessary to point out that even in the higher and best-studied animals we see many structures that are so highly developed that nobody doubts that they are of importance, even though their use has not been, or has only recently been, determined. Mr. Bronn gives the length of the ears and tail in the several species of mice as instances, though trifling ones, of differences in structure that can be of no special use to the owner. However, Dr. Joseph Schöbl, a Czech anatomist, notes that the external ears of the common mouse are supplied in an extraordinary manner with nerves, so that they no doubt serve also as tactile organs; thus the length of the ears can hardly be unimportant. Moreover, as I will presently clarify, we now find that tails are highly useful prehensile organs for the members of some species; thus the tail's use would certainly be influenced by its length.

Since Mr. von Nägeli focuses his arguments on plants, I will do so as well. I certainly admit that the flowers of orchids present a great many curious structures, which a few years ago would have been considered to be mere morphological differences without any special function. But we now know that these odd structures are of the highest importance for achieving fertilization through the aid of insects, and have probably been gained through natural selection. No one until recently would have imagined that in dimorphic and trimorphic plants—plants with either two or three different types of flowers—the different lengths of the stamens (which produce pollen) and pistils (which produce eggs and seeds) (Figure 7.1), and their arrangement on the flower, could possibly have been of any practical use, but we now know that they are. For example, in certain whole groups of plants the ovules stand erect, and in others they are suspended; but in a few other species one ovule stands erect while a second ovule is instead suspended, all within the same ovarium. These positions seem at first to have no functional significance, but the Director of the Royal Botanical Gardens Dr. Joseph Hooker informs me that within the same ovarium, only the upper ovules are fertilized in some cases, and in other cases only the lower ones are fertilized. He suggests that this probably depends on the direction in which the pollen tubes enter the ovarium. If so, then the exact position of the ovules, even when one is erect and the other is suspended within the

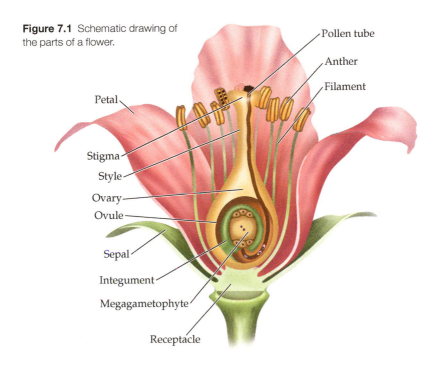

Figure 7.1 Schematic drawing of the parts of a flower.

Pollen tube

Anther

Filament

Petal

Stigma

Style

Ovary

Ovule

Sepal

Integument

Megagametophyte

Receptacle

same ovarium, would follow from the selection of any slight deviations in position that favored fertilization and seed production.

Several plants belonging to distinct orders routinely produce flowers of two kinds—one that is open and with an ordinary structure, and the other that is closed and imperfect. These two kinds of flowers sometimes differ wonderfully in structure and yet may be seen to graduate into each other on the same plant. The ordinary and open flowers can be successfully inter-crossed, with obvious benefits. However, the closed and imperfect flowers are manifestly of great importance as well, as they yield with the utmost safety a large stock of seed with the expenditure of wonderfully little pollen. As just stated, the two kinds of flowers often differ much in structure. The petals in the imperfect flowers almost always consist of mere undeveloped rudiments, and the pollen grains are reduced in diameter. Five of the alternate stamens are rudimentary in the flowering plant *Ononis columnae*; in some species within the violet genus *Viola*, three of the stamens are in this same condition, with two retaining their proper function but being of very small size. In 6 out of 30 of the closed flowers that I examined in an Indian violet (I can't tell the species, as none of the plants ever produced perfect flowers under my care), the sepals[1] were reduced from the normal number of five to only three. In one section of the Malpighiaceae (a family of exclusively tropical and subtropical plants), the closed flowers, according

---

[1] Sepals make up the calyx of flowering plants, which encloses the developing flower bud.

to the French botanist Antoine Laurent de Jussieu, are still further modified, for the five stamens that stand opposite the sepals are all aborted, with a sixth stamen standing opposite a petal being the only one developed; this stamen is not present in the ordinary flowers of these species. Moreover, the style is aborted, and the ovaria are reduced from three to two. Now although natural selection may well have had the power to prevent some of the flowers from expanding and to reduce the amount of pollen—since pollen would be superfluous in a closed flower—yet hardly any of the above special modifications can have been thus determined, but must have followed from what I have called "the laws of growth." These laws would include the functional inactivity of parts as pollen production was gradually reduced and the flowers were opened less and less. Also owing to these rather mysterious "laws of growth," we often see differences in the same parts or organs caused by differences in their relative positions on the same plant. In the Spanish chestnut tree, for example, and in certain fir trees, the leaves on the nearly horizontal branches diverge at different angles from those on the upright branches. And in the British plant *Adoxa*, which gives off a decidedly musk-like scent, the uppermost flower generally has two calyx lobes with the other organs being tetramerous (i.e., divided into fours), while the surrounding flowers generally have three calyx lobes, with the other organs being pentamerous (i.e., divided into five parts). As far as we can tell, such modifications follow from the relative position and interaction of the parts, and not as a result of natural selection.

In many other cases we find modifications of structure that botanists generally consider to be highly important even though they affect only some of the flowers on a given plant, or occur on distinct plants that grow close together under the same conditions. As these variations seem of no special use to the plants, they cannot have been influenced by natural selection. As before we are quite ignorant of the causes of these variations, and can't even invoke any proximate agency, such as their relative position on the organism. Here I will give only a few instances of many that could be given. We commonly see that some flowers on a given plant are four rayed, others three rayed, and so forth. The Swiss botanist Mr. Augustin Pyramus de Candolle, for example, notes that the flowers of the Iranian poppy (*Papaver bracteatum*) offer either two sepals with four petals, which is typical of most poppies, or three sepals with six petals. The French botanist Augustin Saint-Hilaire gives the following similar cases: the genus *Zanthoxylon* belongs to a division of the citrus family Rutaceae with a single ovary; but in some species, flowers with either one or two ovaries may be found on the same plant, and even within the same cluster of flowers (i.e., panicle). Mr. Saint-Hilaire found for *Gomphia olivaeformis* toward the southern extreme of its range, two flower forms of which he was initially certain were distinct species; but he later saw both forms growing on the same bush.

We thus see that with plants, many morphological changes may be governed by factors independent of natural selection. But with respect to Mr.

von Nägeli's doctrine of an innate tendency toward perfection or progressive development, can it be said in the case of these strongly pronounced variations that the plants have been caught in the act of progressing toward a "higher state of development?" On the contrary, the fact that the parts in question differ or vary greatly on the same plant suggests that such modifications are of extremely small importance to the plants themselves, although they may be helpful to us for our classifications. The acquisition of a useless part can hardly be said to raise an organism in the natural scale of complexity; and in the case of the imperfect, closed flowers described above, if any new principle has to be invoked, it must be one of retrogression rather than of progression, as it must be as well with the many parasitic animals that have become secondarily simplified for their new lifestyles.

If the above characteristics are unimportant for the welfare of the species, any slight variations that occurred in these parts would not have been accumulated and augmented through natural selection. When a structure that has been developed through long-continued selection ceases to be of service to a species, it generally becomes variable, for it will no longer be regulated by this same power of selection; we see this with rudimentary organs. And when modifications have been induced that do not affect the success of the species, they may be, and apparently often have been, transmitted in nearly the same state to numerous, otherwise modified descendants. It cannot have been of much importance to the greater number of mammals, birds, or reptiles whether they were clothed with hair, feathers, or scales, and yet hair has been transmitted to almost all mammals, feathers to all birds, and scales to all true reptiles. Even so, any structure that is common to many allied forms is ranked by us as of high importance for classifying organisms, and consequently is often assumed to be of vital importance to the species. I am inclined to believe, however, that many morphological differences that we consider to be important—such as the arrangement of the leaves, the divisions of the flower or of the ovarium, the position of the ovules, and so forth—first appeared in many cases as simple fluctuating variations, which sooner or later became constant not through natural selection, but rather through the nature of the organisms and of the surrounding conditions, as well as through the interbreeding of distinct individuals. For as these morphological characteristics do not affect the welfare of the species, any slight deviations in them could not have been governed or accumulated through selection. It is a strange result that we thus arrive at: characteristics that are of the most importance to the systematist are only of slight importance to the lives of the organisms they study. Later, when I talk about the genetic principle of classification (Chapter 14),[2] we will see that this is by no means as paradoxical as it may at first appear.

Although we have no good evidence of the existence in living organisms of any innate tendency towards progressive development, such a tendency

---

[2] Darwin's cross-references to chapters beyond Chapter 8 have been retained in this volume so readers can refer to the original *The Origin of Species*, Sixth Edition.

would not necessarily be expected to follow from the actions of natural selection, as I have tried to show in Chapter 4. Indeed, the best definition that has ever been given of a high standard of organization is the degree to which various body parts have been specialized or differentiated; natural selection tends toward specialization, inasmuch as the parts are thus enabled to perform their functions more efficiently.

Supposed Incompetence of Natural Selection to Account for the Incipient Stages of Useful Structures

The distinguished zoologist Mr. St. George Mivart has recently collected all the objections that have been advanced to date against the theory of natural selection, as put forward by both Mr. Alfred Russel Wallace and myself, and has illustrated them with admirable art and force. When thus marshaled, they make a formidable array. However, as Mr. Mivart does not give any of the various facts and considerations that go against his conclusions, readers have no chance to weigh the evidence on both sides. When discussing special cases, Mr. Mivart passes over the effects of the increased use and disuse of parts, which I have always maintained to be highly important, and which I have addressed in my book *The Variation of Animals and Plants Under Domestication* at, I believe, greater length than any other writer. He likewise often assumes that I attribute nothing to variation independently of natural selection, whereas in *The Variation of Animals and Plants Under Domestication* I have collected more well-established cases than can be found in any other work that I know of. My judgment may not be trustworthy, but after reading Mr. Mivart's book with considerable care, and comparing each section with what I have said on the same topic, I never before felt so strongly convinced of the general truth of the conclusions that I have arrived at here.

All of Mr. Mivart's objections will be, or have already been, considered in this book. The one new point that appears to have struck many readers is, "that natural selection is incompetent to account for the incipient stages of useful structures." This subject is intimately connected with that of the gradation of characteristics, often accompanied by a change of function—for instance, the conversion of a swim bladder into lungs—points that I discussed at length in the previous chapter under two headings. Nevertheless, I will here consider in some detail several of the cases advanced by Mr. Mivart, selecting those that are the most illustrative, as want of space prevents me from considering them all. In every case I think you will agree that natural selection prevails in accounting for what we see among organisms in nature today.

First let us consider the giraffe. This animal, by its lofty stature and its much elongated neck, forelegs, head, and tongue, has its whole frame beautifully adapted for browsing on the higher branches of trees. It can thus obtain food well beyond the reach of the other hoofed animals (i.e., ungulates) inhabiting the same lands; this must be a great advantage to it during times of famine. Indeed, the Niata cattle in South America show us how important a small change in structure could be in preserving an animal's life during such periods. These cattle can browse on grass as well as other animals do, but during droughts, because of the way their lower jaw projects, they cannot browse on the twigs of trees, reeds, and so forth on which the common cattle and horses are driven and able to feed. At such times, without the ability to feed on other foods, and unless they are fed by their owners, the Niata cattle perish in large numbers.

Before coming to Mr. Mivart's specific objections, let me just summarize once again how natural selection will act in all ordinary cases. We have modified some of our domestic animals without having necessarily paid much attention to special structural features, simply by preserving and breeding from the fleetest individuals, as with the racehorse and the greyhound, or by simply preserving and breeding solely from the victorious birds after staged fights, as with the gamecock. Likewise, with the ancestral giraffe in nature, those individuals that did their browsing higher up than others, and were able during famines to reach even an inch or two above the others, will often have been more likely to feed better and thus survive and reproduce as they roamed over the whole country in search of food. That individuals of the same species often differ slightly in the relative lengths of all their parts may be seen in the many works of natural history in which careful measurements are given. Such differences will have been of great importance to the ancestral giraffe, considering its probable habits of life; for those individuals that had some one of several body parts rather more elongated than usual would generally have been most likely to survive when food was scarce. These individuals will have been more likely to have mated and left offspring, which then would either have inherited the same body peculiarities or at least had a tendency to vary again in the same manner. In contrast, those individuals that were less favored in the same respects will have been most likely to have perished.

Note that when we methodically try to improve a breed, we separate all the superior individuals, allowing only those individuals to freely interbreed. In contrast, natural selection will preserve all the superior individuals, allowing them to interbreed freely, and will destroy all the inferior individuals. This process corresponds exactly with what I have called unconscious selection by man. Continuing the process in nature for many, many generations could easily, it seems to me, convert an ordinary hoofed quadruped[3] into a giraffe.

Mr. Mivart offers two objections to this conclusion. One is that an increased body size would obviously require an increased supply of food; he suggests that it is "very problematical whether the disadvantages thence arising would not, in times of scarcity, more than counterbalance

---

[3] Any four-legged animal, such as dogs, horses, and cows.

the advantages." But as giraffes do in fact exist in large numbers in Africa, and as some of the largest antelopes in the world—taller in fact than an ox—abound there, why should we doubt that, as far as size is concerned, intermediate gradations in height could formerly have existed there, even when subjected as now to severe famine? Surely, being able to reach, at each increase in size, a supply of food unreachable by other hoofed quadrupeds in the area would have been of some advantage to the nascent giraffe. In addition, the increased bulk of the animals would help to protect them from almost all predators other than the lion; and against that animal, the ancestral giraffe's increasingly tall neck—and the taller the better—would, as the American philosopher and mathematician Mr. Chauncey Wright has remarked, serve as a valuable watchtower for seeing potential predators from afar. Indeed, as the British explorer and naturalist Sir Samuel White Baker remarks, no animal is more difficult to stalk than a giraffe. This animal also uses its long neck as a means of offense or defense, by violently swinging its head—which is armed with stump-like horns—back and forth. The preservation of each species will rarely be controlled by any one advantage, but rather by the combination of them all, both great and small.

Mr. Mivart's second objection here is that if natural selection is so potent, and if browsing on vegetation higher up is such a great advantage, then why haven't any other hoofed quadrupeds acquired a long neck and lofty stature other than the giraffe, and to a lesser extent the camel, guanaco (Figure 7.2A), and the now-extinct, long-necked llama, *Macrauchenia* (Figure 7.2B)? Or again, why haven't any members of the group acquired a long proboscis? The answer is not difficult when considering South Africa, which was formerly inhabited by numerous herds of giraffe, and can best be given through the following illustration. In every meadow in England in which trees grow, we see the lower branches trimmed to an exact level by the browsing of horses and cattle; so what advantage could there be for sheep, for example, living in that same area, to acquire slightly longer necks when there

(A)   (B)

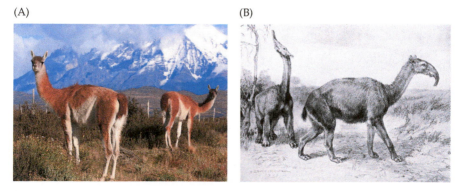

Figure 7.2 (A) Guanacos (*Lama glama*). (B) South American llamas (*Macrauchenia patachonica*), which went extinct 10,000–20,000 years ago.

is no food for them to eat higher up? In every district, some one kind of animal will almost certainly be able to browse higher than the others; and it is almost equally certain that this one kind alone could have its neck elongated for this purpose, through natural selection and the effects of increased use. In South Africa, the competition for browsing on the higher branches of the acacias (Figure 7.3) and other trees must be between giraffe and giraffe, and not with the other ungulate animals.

Figure 7.3 Giraffes (*Giraffa camelopardalis*) eating from an acacia tree.

We don't know why various animals belonging to this same category of animals in other parts of the world have not acquired either an elongated neck or a proboscis. But it is as unreasonable to expect a distinct answer to such a question as to ask why some event in the history of humankind occurred in one country but not in another. We don't know what conditions determine the numbers and range of each species; and we cannot even guess what anatomical changes would favor the increase of any species in some new country. We can, however, see in a general manner that various causes might have interfered with the development of a long neck or proboscis in some animals. To reach the foliage at a considerable height implies a substantial increase in the bulk of the body, unless of course the animals can climb trees, something that hoofed animals are particularly ill-equipped to do. We know that some very luxuriant areas, such as South America, support remarkably few large quadrupeds, while South Africa abounds with them to an unparalleled degree. Why this should be so, we don't know. Nor do we know why the later Tertiary period, which ended about 66 million years ago, should have been so much more favorable for their existence than the present time. Whatever the causes might have been, we can see that certain regions and times would have been much more favorable than others for the development of so large a quadruped as a giraffe.

It seems almost certain that an animal cannot acquire some special, well-developed anatomical modification without also having several other body parts modified and coadapted in some way. Although every part of the body varies slightly among individuals, it does not follow that the necessary body parts should always vary in the right direction and to the right extent. With the different species of our domesticated animals we know that the various parts vary in different ways and to different degrees, and that some species show more variability than others. But even if suitable variations arise, it does not follow that natural selection would always be able to act on them to produce a structure that would be clearly beneficial to the species. For instance, if the number of individuals existing in a region is determined chiefly by the extent of predation or by the impact of internal or external parasites, as often seems to be the case, then natural selection will be able to do little in modifying any particular structure for obtaining food.

Lastly, natural selection is a slow process, and the same favorable conditions must endure for a considerable time before any marked effect can be produced. These reasons are very general and very vague, but that's all we can say at the moment about why, in many quarters of the world, most hoofed quadrupeds have not acquired greatly elongated necks or other means for browsing on the higher branches of trees.

Objections similar to those just mentioned have also been advanced by many other writers. In each case, a variety of factors, in addition to the general ones just discussed, have probably prevented the gradual acquisition of particular structures through natural selection, even though such modifications would seem useful for those organisms to have had. One writer asks, "Why has not the ostrich acquired the power of flight?" But think about the enormous supply of food that would be needed to give this desert bird the ability to move its huge body through the air. Similarly, oceanic islands are inhabited by bats and seals, but not by any terrestrial animals. Yet some of those bats are very distinctive, indicating that they must have occupied those islands for a very long time. Thus, Sir Charles Lyell asks, Why haven't seals and bats given birth on such islands to forms that can live well on land? But seals would first have to be converted into terrestrial carnivores of a considerable size, and bats into terrestrial insectivorous animals. That would be a great problem, since for seals there would be nothing for them to prey on, and for the bats, the ground insects that would be available for food would already be largely preyed on by the reptiles and birds that first colonized and now abound on most oceanic islands.

Gradations of structure, with each stage beneficial to a changing species, will be favored only under certain peculiar conditions. A strictly terrestrial animal, by occasionally hunting for food in shallow water, then in streams or lakes, might eventually be converted into an animal so thoroughly aquatic as to brave the open ocean. But seals would not find the conditions on oceanic islands favorable for their gradual reconversion into a terrestrial form. Bats, as formerly shown, probably acquired their wings by at first gliding through the air from tree to tree, like the so-called flying squirrels (see Figure 6.1C) of today, for the sake of escaping from their enemies, or for avoiding falls. But when the power of true flight had once been acquired it would never be reconverted back, at least not for the above purposes, into the less efficient power of simply gliding through the air. Bats might, indeed, like many birds, have had their wings greatly reduced in size, or completely lost through disuse. In such a case, however, they would first have had to acquire the ability to run quickly on the ground using their hind legs alone, so as to compete successfully with birds or other ground animals; bats seem singularly ill-fitted for such a change.

I have made these conjectural remarks only to show that a transition of structure, with each step beneficial to its owner, is a highly complex affair, and that there is nothing strange in a transition not having occurred in any particular case.

Finally, more than one writer has asked, Why have some animals had their mental powers more highly developed than others? Wouldn't such development be advantageous to all? Why haven't apes acquired the intellectual powers of man? Various explanations could be given, but they are all conjectural and their relative probabilities cannot be weighed. A definite answer to the question should not be expected, seeing that no one can solve the simpler problem of why, of two races of primitive peoples, one has risen higher in the scale of civilization than the other, something that presumably implies increased brainpower.

Let us return to more of Mr. Mivart's objections. Insects often resemble various other objects for the sake of protection—green or decayed leaves, for example, or dead twigs, bits of lichen, flowers, spines, and even the excrement of birds. The resemblance is often wonderfully close, not just in color but in form as well, and even to the manner in which the insects hold themselves. The caterpillars that stick out motionless from the bushes on which they feed, like dead twigs, offer an excellent example of a resemblance of this kind. Cases in which insects imitate the excrement of birds are rare and exceptional. On this point, Mr. Mivart remarks, "As, according to Mr. Darwin's theory, there is a constant tendency to indefinite variation, and as the minute incipient variations will be in *all directions*, they must tend to neutralize each other, and at first to form such unstable modifications that it is difficult, if not impossible, to see how such indefinite oscillations or infinitesimal beginnings can ever build up a sufficiently appreciable resemblance to a leaf, bamboo, or other object for natural selection to seize upon and perpetuate."

But in all the foregoing cases, the insects in their original state no doubt presented some rude and accidental resemblance to an object commonly found in the habitats that they frequented. Nor is this at all improbable, considering the almost infinite number of surrounding objects and the diversity in form and color of the various existing insects. As some rude resemblance is necessary for the first start, we can understand how it is that the larger and higher animals do not (with the single exception, as far as I know, of one fish) gain protection by resembling special objects, but only the surface that commonly surrounds them, and this chiefly through color. Assuming that an insect originally happened to resemble in some degree a dead twig or a decayed leaf, and that it varied slightly in many ways, then all the variations that rendered the insect at all more like any such object, and thus favored its escape from predation, would be preserved, while other

Figure 7.4 A walking stick insect.

variations would be neglected and ultimately lost. Or, if they rendered the insect at all less like the imitated object, they would be eliminated through predation. Mr. Mivart's objection would indeed be valid if we were attempting to account for the above resemblances independently of natural selection, through mere fluctuating variability, but as that is not at all what we are trying to do, there is no problem to be resolved.

Nor can I see any force in Mr. Mivart's difficulty with respect to "the last touches of perfection in the mimicry," as in the case given by Mr. Wallace of a walking stick insect (Figure 7.4) found on Borneo that resembles "a stick grown over by a creeping moss or jungermannia." So close was this resemblance in fact, that a Borneo native maintained that the foliaceous excrescences (leafy outgrowths) really were moss! Insects are preyed on by birds and other enemies, whose sight is probably sharper than ours, and every grade in resemblance that aided an insect in escaping notice or detection would help its survival; the more perfect the resemblance, the safer the insect. Considering the nature of the differences between the species in the group that includes the above-mentioned stick insect, there is nothing improbable in this insect showing variations in the irregularities on its surface, and in those irregularities having become more or less green-colored over the generations; for within every group, the characteristics that differ among the several species are the most apt to vary, while the generic characteristics, or those common to all the species in the group, are the most constant.

The Greenland whale is one of the most wonderful animals in the world, and the baleen (Figure 7.5) or whalebone, one of its greatest peculiarities. The baleen consists of a row of about 300 thin plates ("laminae") on each side of the upper jaw. These plates stand close together transversely to the longer axis of the mouth, and within the main row there are some subsidiary rows. The extremities and inner margins of all the plates are frayed into stiff bristles that clothe the whole gigantic palate and serve to strain or sift particles from the water, and thus to concentrate and retain the minute prey on which these huge animals subsist. The middle and longest lamina in the Greenland whale is 10, 12, or even 15 feet long. In other cetacean species we also see graduations in length: in one species, the middle lamina is 4 feet long, according to the Arctic explorer William Scoresby, in another it is 3 feet long, in another 18 inches long, and in *Balaenoptera acutorostrata*, one of the Rorqual or Minke whales, only about 9 inches long. The quality of the whalebone also differs among the different species.

(A)

(B)

Figure 7.5 (A) A whale's open mouth showing baleen. (B) One plate of baleen from a baleen whale.

With respect to the baleen itself, Mr. Mivart remarks that once it "had attained such a size and development as to be at all useful, then its preservation and augmentation within serviceable limits would be promoted by natural selection alone. But how to obtain the beginning of such useful development?" In answer, can't we imagine that the early ancestors of the baleen whales possessed a mouth constructed something like the lamellated beak of a duck,[4] with its rows of comb-like structures ("pectin") along the beak? Ducks, like whales, subsist by sifting the mud and water for food; indeed, the family has sometimes been called Criblatores, or "sifters." I hope I may not be misconstrued into saying that the progenitors of whales actually did possess mouths lamellated like the beaks of ducks. I only wish to show that this is not incredible, and that the immense plates of baleen seen in the Greenland whale might have been developed from such lamellae by finely graded steps, each of use to its owner.

Indeed, the beak of a shoveler duck (*Anas clypeata*) (Figure 7.6A) is a more beautiful and complex structure than the mouth of a whale. In the specimen that I examined, the upper mandible is furnished with a row of 188 thin, elastic lamellae on each side, obliquely beveled so as to be pointed, and placed transversely to the longer axis of the mouth. They arise from the palate and are attached to the sides of the mandible by flexible membranes. Those lamellae standing towards the middle are the longest, being about one-third of an inch long, and they project fourteen-hundredths of an inch beneath the edge. At their bases there is a short subsidiary row of obliquely transverse lamellae. In these several respects they resemble the plates of baleen in the mouth of a whale. But toward the end of the beak they differ quite a bit, as they project inward instead of straight downward. The entire head of the shoveler, though incomparably less bulky, is about 1/18 of the length

---

[4] Some duck beaks have a series of tooth-like plates or ridges (see Figure 7.6), used for sieving food particles from the water.

(A)

(B)

(C)

(D)

(E)

Figure 7.6 (A) A male shoveler duck (*Anas clypeata*), showing baleen-like projections. (B) Two male torrent ducks (*Merganetta armata*). (C) A male wood duck (*Aix sponsa*). (D) The Egyptian goose (*Alopchen aegyptiacus*). (E) A male merganser, showing its serrated beak (*Mergus merganser*).

of the head of a moderately large *Balaenoptera acutorostrata*, the Minke whale, in which species the baleen is only 9 inches long as noted earlier. Thus, if we were to make the head of the shoveler duck as long as that of *Balaenoptera*, the lamella would be 6 inches long—that is, two-thirds of the length of the baleen found in this species of whale. The lower mandible of the shoveler duck is furnished with lamellae equal in length with those above, but finer; and in being thus furnished it differs conspicuously from the lower jaw of a whale, which is destitute of baleen. On the other hand, the extremities of this duck's lower lamellae are frayed into fine bristly points, curiously resembling the plates of baleen. In the bird genus *Prion*, a member of the distinct family of petrels, only the upper mandible is furnished with lamellae. These are well developed and project beneath the margin; thus the beak of this bird resembles, in this respect, the mouth of a whale.

From the highly developed structure of the shoveler's beak we may then proceed, without any great break as far as fitness for sifting is concerned, through the beak of the torrent duck (*Merganetta armata*) (Figure 7.6B), and in some respects through that of the wood duck (*Aix sponsa*) (Figure 7.6C) to the beak of the common duck. I am basing my remarks on information and specimens sent to me by the ornithologist Mr. Osbert Salvin. The lamellae found in the common duck are much coarser than those of the shoveler, and are firmly attached to the sides of the mandible. There are only about 50 on each side, and they do not project at all beneath the margin. They are all square-topped, and are edged with translucent hardish tissue, as if for crushing food. The edges of the lower mandible are crossed by many fine ridges that project very little. Compared with the beak of the shoveler, the common duck's beak is thus very inferior for sifting, and yet this bird, as everyone knows, constantly uses it for exactly this purpose! There are other species, as I hear from Mr. Salvin, in which the lamellae are considerably less developed than in the common duck; but I do not know whether or not the members of those species use their beaks for sifting food from the water.

Let us turn now to another group of the same family. The beak of the Egyptian goose (*Alopochen*) (Figure 7.6D) closely resembles that of the common duck, except that the lamellae are not so numerous and not so distinct from each other, and do not project so much inward. Yet this goose, I am told by the English ornithologist Mr. Edward Bartlett, "uses its bill like a duck by throwing the water out at the corners." Its chief food, however, is grass, which it crops like the common goose does. In that latter bird (the common goose), the lamellae of the upper mandible are much coarser than they are in the common duck and almost confluent, with about 27 on each side, and they terminate upward in teeth-like knobs. The palate is also covered with hard rounded knobs. The edges of the lower mandible are serrated with teeth that are much more prominent, coarser, and sharper than those of the duck. The common goose does not sift the water at all, but instead uses its beak exclusively for tearing or cutting herbage, for which purpose it is so well fitted that it can crop grass closer to the ground than almost any other animal. There are other species of geese, as I hear from Mr. Bartlett, in which the lamellae are less developed than those of the common goose.

Thus we see that a member of the duck family, with a beak constructed like that of the common goose and adapted solely for grazing, or even a member with a beak having less well-developed lamellae, might be converted by small changes into a species like the Egyptian goose, and this into one like the common duck, and this, lastly, into one like the shoveler, a bird that is provided with a beak almost exclusively adapted for sifting food from the water; indeed, this bird is so specialized that it could hardly use any part of its beak, except the hooked tip, for seizing or tearing solid food. Let me also add that a goose's beak might be converted by small gradual changes into one with prominent, recurved (curved backward or inward) teeth, like those of the merganser (*Mergus merganser*) (Figure 7.6E), which belongs to the same family.

Returning to the whales now, the northern bottlenose whale (*Hyperoodon ampullatus*, formerly *H. bidens*) (Figure 7.7) lacks true, functional teeth, according to the French naturalist Bernard Germain de Lacépède, but its palate is roughened with small, unequal, hard points of horn. There is, therefore, nothing improbable in supposing that some ancestral cetacean was provided with similar but more regularly spaced points of horn on

Figure 7.7 Northern bottle-nose whale (*Hyperoodon ampullatus*).

its palate that aided it in seizing or tearing its food, just as the knobs on a goose's beak do now. If so, it will hardly be denied that the points might then have gradually been converted through variation and natural selection into lamellae that are as well developed as those of the Egyptian goose; in that case they would have been used both for seizing objects and for sifting food from the water. Then we can imagine them being gradually converted into lamellae like those of the domestic duck, and from there onward until they were as well constructed as those of the shoveler, in which case they would have served exclusively as a sifting apparatus. From this stage, in which the lamellae would be two-thirds the length of the plates of baleen found in *Balaenoptera rostrata*, gradations—which may still be observed in some modern cetaceans—lead us onward to the enormous plates of baleen now seen in the Greenland whale. Nor is there the least reason to doubt that each step in this scale might have been as useful to certain ancient cetaceans, with the functions of the parts slowly changing over the generations, as the various gradations in the beaks of the different modern members of the duck family are now. We should bear in mind that each species of duck is subjected to a severe struggle for existence, so that the structure of every part of its frame must be well adapted to its conditions of life.

The flatfish, members of the fish family Pleuronectidae (Figure 7.8A), are remarkable for their asymmetrical bodies. They rest on one side—usually the left—and occasionally we find reversed adult specimens. At first sight the lower, resting surface resembles the ventral surface of an ordinary fish: it is white in color, less developed in many ways than the upper side, and with the lateral fins often of smaller size. But the eyes offer the most remarkable peculiarity in adults: both eyes are found on the upper side of the head! During the juvenile stage, however, the eyes are positioned on opposite sides of the head just like those of a normal fish; moreover, both sides of the juvenile's body are pigmented equally. As development proceeds, however, the eye that is located on the side of the body that will eventually be facing downwards against the substrate begins to glide slowly around the head to the upper side of the body. Obviously, if this lower eye did *not* move to the other side of the

(A)   (B)

Figure 7.8 (A) A flatfish. Note that both eyes are on the same side of the head. (B) The American plaice (*Hippoglossoides platessoides*).

body during development, it would end up pointing into the substrate and would in fact be abraded by the sandy bottom. That flatfish are admirably adapted by their flattened and asymmetrical structure for their habits of life is evident from the fact that the various flatfish species (soles, flounders, and halibut, for example) are extremely common. The chief advantages thus gained seem to be protection from their enemies and facility for feeding on the sediment. However, the different members of the family present, as the Danish biologist Jørgen Matthias Christian Schiødte remarks, "a long series of forms exhibiting a gradual transition from *Hippoglossus pinguis*, which does not in any considerable degree alter the shape in which it leaves the ovum, to the soles, which are entirely thrown to one side."

Mr. Mivart, in considering this case, remarks that a sudden spontaneous transformation in the position of the eyes is hardly conceivable, in which I quite agree with him. He then adds. "If the transit was gradual, then how such transit of one eye a minute fraction of the journey toward the other side of the head could benefit the individual is, indeed, far from clear. It seems, even, that such an incipient transformation must rather have been injurious." But he might have found a satisfying answer to this objection in the excellent observations published in 1867 by the Swedish zoologist August Wilhelm Malm. Flatfish, it turns out, when very young and still fully symmetrical, and with their eyes located on opposite sides of the head, cannot long retain a vertical position, owing to the excessive depth of their bodies, the small size of their lateral fins, and to their lacking a swim bladder. Thus, soon growing tired, they fall to the bottom on one side. Mr. Malm has observed that while the fish are thus resting, they often twist the lower eye upward, trying to see above them. Indeed, they do this so vigorously that the eye is pressed hard against the upper part of the orbit. The forehead between the eyes consequently becomes temporarily contracted in breadth. On one occasion Mr. Malm actually saw a young fish raise and depress the lower eye through an angular distance of about 70 degrees.

We should remember that at this early age the skull is cartilaginous and flexible, so that it readily yields to muscular action. We also know that the skull of higher animals can be altered in shape if the skin or the muscles are permanently contracted through disease or some accident, even in young adulthood. Indeed, with long-eared rabbits, if one ear lops forward and downward, its weight drags forward all the bones of the skull on the same side. Mr. Malm states that the newly hatched young of perch, salmon, and several other symmetrical fishes have the habit of occasionally resting on one side at the bottom; of considerable interest to us, he has observed that they often then strain their lower eyes so as to look upwards, so that their skulls are thus rendered rather crooked. These fishes, however, are soon able to hold themselves in a vertical position, so that no permanent effect is thus produced.

Mr. Schiødte believes, in opposition to some other naturalists, that members of the family Pleuronectidae (righteye flounders) are not quite symmetrical even as embryos; if so, we could understand how it is that certain species, while young, habitually fall over and rest on their left side, and other species on their right side. Mr. Malm adds, in confirmation of this view, that adults of the dealfish *Trachipterus arcticus* (a type of ribbonfish), which is not a member of the family Pleuronectidae, rests at the bottom on its left side and swims diagonally through the water; in this fish, the two sides of the head are said to be somewhat dissimilar. Our great authority on fishes, Dr. Albert Günther, concludes his abstract of Mr. Malm's paper by remarking that "the author gives a very simple explanation of the abnormal condition of the Pleuronectoids."

We thus see that the first stages of the transit of the eye from one side of the head to the other, which Mr. Mivart assumes would be injurious, may be attributed to the habit of endeavoring to look upward with both eyes while resting on one side at the sea's bottom—something that no doubt benefits both the individual and the species. We may also attribute to the inherited effects of use the fact that the mouth in several kinds of flatfish is shifted toward the lower surface, with the jaw bones stronger and more effective on this side of the head—the eyeless side—probably, as the Scottish flatfish authority Dr. Ramsay Traquair supposes, for feeding with ease on the bottom. On the other hand, disuse[5] probably accounts for the less developed condition of the whole inferior half of the body, including the lateral fins, though the English zoologist William Yarrell thinks that the reduced size of these fins is advantageous to the fish, as "there is so much less room for their action, than with the larger fins above." Perhaps the fact that there are 25–30 teeth in the lower halves of the two jaws of the plaice (Figure 7.8B), but only 4–7 in the upper halves, may likewise be accounted for by disuse. From the colorless state of the ventral surface of most fishes and of many other animals, we may reasonably suppose that the absence of color in flatfish on whichever side faces downward is due to the exclusion of light. But

---

[5] As we saw earlier, particularly in Chapter 5, Darwin thought—erroneously, as it turns out—that the effects of use and disuse in the parent could in some cases be directly inherited by offspring.

it cannot be supposed that the action of light is responsible for the peculiar speckled appearance of the upper side of the sole, so closely resembling the sandy bed of the sea, or for the ability of individuals of some species to alter their body color to closely match that of the surrounding surface (as recently shown by Charles Henri Georges Pouchet at the Muséum National d'Histoire Naturelle in Paris), or for the presence of bony tubercles (nodules) on the upper side of the turbot. Instead, natural selection has probably come into play here, as it probably has in adapting the general shape of the body of these fishes, and many other peculiarities, to their habits of life. We should keep in mind, as I have insisted already several times, that the inherited effects of the increased use of parts, and perhaps of their disuse, will be strengthened by natural selection; all spontaneous variations in the right direction will thus be preserved, as will individuals who strongly inherit any beneficial effects of increased use. How much to attribute in each particular case to the effects of use and disuse, and how much to natural selection, it seems impossible to decide at present.

Let us now consider the mammals. By definition, all mammals have mammary glands, which are indispensable to their existence. Such glands must, therefore, have been developed a very long time ago, and we can know nothing definite about the details of their origins. Mr. Mivart asks: "Is it conceivable that the young of any animal was ever saved from destruction by accidentally sucking a drop of scarcely nutritious fluid from an accidentally hypertrophied cutaneous gland[6] of its mother? And even if one was so, what chance was there of the perpetuation of such a variation?" But Mr. Mivart does not put the case fairly. Most evolutionists admit that mammals are descended from a marsupial form. If so, the mammary glands will have initially developed within the marsupial sack. Even in the case of seahorses (members of the fish genus *Hippocampus*) (Figure 7.9), the eggs are hatched—and the young are then reared for a time—within a sack of exactly this nature. Indeed, an American naturalist, the Reverend Dr. Samuel Lockwood, believes from what he has seen of the development of the young that they are nourished by a secretion from the cutaneous glands of that sack. Now with the early mammalian ancestors, almost before they deserved to be thus designated, is it not at least possible that the young might have been similarly nourished? And in this case, any individuals that secreted a fluid that was in some degree or manner especially nutritious, so as to take on some of the nutritious characteristics of milk, would

Figure 7.9 A seahorse (*Hippocampus* sp.).

---

[6] An increased volume of a gland in the skin due to an increase in the size of its component cells.

in the long run have reared a larger number of well-nourished offspring than would individuals that secreted a fluid poorer in nutrients. Thus the cutaneous glands, which are the homologues[7] of the mammary glands, would have been improved and made more effective over the generations. It fits well with the widely extended principle of specialization, that some glands over a certain space of the sack should have become more highly developed than the rest; they would then have formed a breast, but at first one without a nipple, as indeed we see in the platypus (*Ornithorhynchus anatinus*) (see Figure 4.9A), an animal that sits at the base of mammalian evolution. The development of the mammary glands would have been of no use, and thus could not have been achieved through natural selection, unless the young at the same time were able to feed on the secretion. Understanding how young mammals have instinctively learned to suck their mother's breast is no harder than understanding how unhatched chickens have learned to break the eggshell that encloses them by tapping against it with their specially adapted beaks, or how a few hours after leaving the eggshell they have learned to pick up grains of food. In such cases, the most probable solution seems to be that the habit was first acquired by practice at a more advanced age, and afterward transmitted to the offspring earlier in their development. But the young kangaroo is said not to suck, only to cling to the nipple of its mother, who is then able to actively inject milk into the mouth of her helpless, half-formed offspring. On this point, Mr. Mivart remarks that without some special provision, "the young one must infallibly be choked by the intrusion of the milk into the windpipe." Mr. Mivart then goes on to note that indeed, there is a special provision: The infant kangaroo's larynx "is so elongated that it rises up into the posterior end of the nasal passage, and is thus enabled to give free entrance to the air for the lungs while the milk passes harmlessly on each side of this elongated larynx, and so safely reaches the gullet behind it." Mr. Mivart then asks, how did natural selection remove in the adult kangaroo (and in most other mammals, on the assumption that they have descended over time from a marsupial form), "this at least perfectly innocent and harmless structure?" Let me suggest that the voice, which is certainly of high importance to many animals, could hardly have been used with full force as long as the larynx entered the nasal passage. Indeed, the well-known comparative anatomist Sir William Henry Flower has suggested to me that this structure would have greatly interfered with an animal swallowing solid food.

Let us now turn to the lower divisions of the animal kingdom. Members of the Echinodermata (the phylum containing such animals as sea stars, sea urchins, and brittle stars) are furnished with remarkable organs called pedicellariae (Figure 7.10A), which consist, when well-developed, of a three-part forceps—that is, each pedicellaria is formed of three serrated,

---

[7] Homologues are parts that have a common ancestry. Here Darwin is suggesting that the mammary glands have their evolutionary origins in these cutaneous glands.

(A)

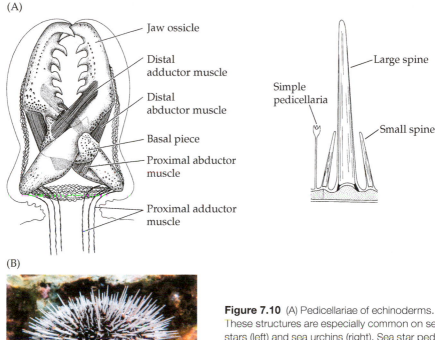

Jaw ossicle

Distal
adductor muscle

Distal
abductor muscle

Basal piece

Proximal abductor
muscle

Proximal adductor
muscle

Large spine

Simple
pedicellaria

Small spine

(B)

Figure 7.10 (A) Pedicellariae of echinoderms.
These structures are especially common on sea
stars (left) and sea urchins (right). Sea star pedicellariae are usually two-jawed whereas urchin
pedicellariae are usually three-jawed.
(B) *Echinus* sp., a member of the animal phylum
Echinodermata.

moveable fingers, neatly fitting together and placed at the top of a flexible
stem, which is moved by muscles. These forceps can firmly seize hold of
any object: The comparative anatomist Louis Agassiz at Harvard University has even seen a sea urchin rapidly passing particles of excrement from
forceps to forceps down certain lines of its body, in order to keep its shell
from being fouled. But there is no doubt that besides removing dirt of all
kinds from the outside of the shell, they also perform other functions, one
of which, apparently, is defense.

With respect to these remarkable structures, Mr. Mivart, as on so many
previous occasions, asks, "What would be the utility of the first *rudimentary
beginnings* of such structures, and how could such incipient buddings have
ever preserved the life of a single *Echinus* [sea urchin]?" He adds, "Not even
the *sudden* development of the snapping action could have been beneficial
without the freely moveable stalk, nor could the latter have been efficient
without the snapping jaws, yet no minute merely indefinite variations could
simultaneously evolve these complex coordinations of structure; to deny
this seems to do no less than to affirm a startling paradox."

Well, as paradoxical as this may appear to Mr. Mivart, the members of
some sea star species in fact do have three-part forceps that are immovably

fixed at the base but that are nevertheless capable of a snapping action; this makes sense if they serve, at least in part, as a means of defense. Mr. Agassiz, to whose great kindness I am indebted for much information on the subject, informs me that there are other sea stars in which one of the three fingers of the forceps is reduced to a support for the other two fingers, and that in some genera the third finger is completely lost. In *Echinoneus*, the shell is described by the French echinoderm expert Jean Octave Edmond Perrier as bearing two kinds of pedicellariae, one resembling those of *Echinus* (Figure 7.10B), and the other those of the heart urchin *Spatangus*. Such cases are always interesting, as they afford the means of apparently sudden transitions, through abortion of one of the two states of an organ.

With respect to the steps through which these curious organs have been evolved, Mr. Agassiz infers from his own research and that of Johannes Peter Müller that the pedicellariae of both sea stars and sea urchins must be looked at as modified spines. This may be inferred from their manner of embryological development, as well as from a long and perfect series of gradations from simple granules to ordinary spines in different species and genera, and then to perfect three-pronged pedicellariae. The gradations extend even to the manner in which ordinary spines and pedicellariae, with their supporting calcareous rods, are articulated to the shell. In certain sea star genera, "the very combinations needed to show that the pedicellariae are only modified branching spines" may be found. Thus we have fixed spines, with three equidistant, serrated, moveable branches articulated near their bases, and higher up, on the same spine, three other moveable branches. Now when the latter arise from the tip of a spine they form in fact a crude three-pronged pedicellaria, and such may be seen on the same spine together with the three lower branches. In this case the similarity between the fingers of the pedicellariae and the moveable branches of a spine is unmistakable. Since it is generally admitted that the ordinary spines serve for protection there is no reason to doubt that those furnished with serrated and moveable branches are also protective; indeed they would offer particularly effective protection if the three branches began to act as a prehensile or snapping apparatus. Thus every gradation, from an ordinary fixed spine to a fixed, snapping pedicellaria, would be of use to its possessor.

In some sea star genera, the pedicellariae are not fixed or borne on an immoveable support, but instead are placed at the top of a short but flexible and muscular stem; in this case they probably perform some other function in addition to defense. In sea urchins we can follow the steps by which a fixed spine becomes articulated to the shell, and thus becomes moveable. I wish I had space here to give a more detailed summary of Mr. Agassiz's interesting observations on the development of pedicellariae. He notes that all possible gradations may also be found between the pedicellariae of sea stars and the hooks found on brittle stars, another group in the phylum Echinodermata, and also between the pedicellariae of sea urchins and the anchors of sea cucumbers (class Holothuroidea, another important echinoderm group).

Widely Different Organs in Members of the Same Class, Developed from One and the Same Source

One very interesting group of colonial animals, namely the Bryozoa (also called Polyzoa and Ectoprocta), are provided with curious organs called avicularia[8] (Figure 7.11A). These differ considerably in structure among the various bryozoan species. In their most perfect condition they curiously resemble the head and beak of a vulture in miniature, seated on a neck and capable of moving, as is the lower jaw as well. In one bryozoan species that I looked at, all the avicularia on the same branch often moved simultaneously backward and forward through an angle of about 90 degrees in the

---

[8] These are small, stalked structures that are specialized for pinching small prey or discouraging larvae of other species from attaching.

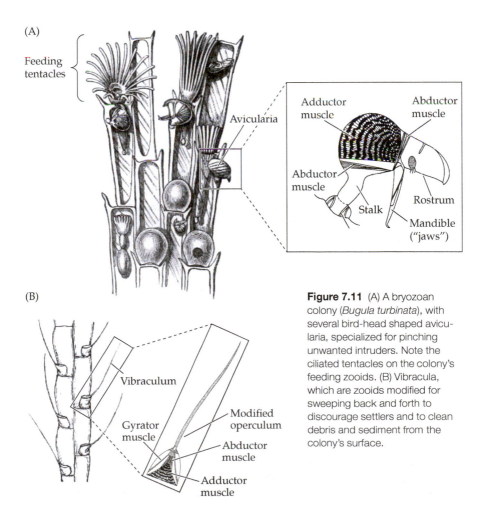

Figure 7.11 (A) A bryozoan colony (*Bugula turbinata*), with several bird-head shaped avicularia, specialized for pinching unwanted intruders. Note the ciliated tentacles on the colony's feeding zooids. (B) Vibracula, which are zooids modified for sweeping back and forth to discourage settlers and to clean debris and sediment from the colony's surface.

course of only 5 seconds, with the lower jaw wide open. When the jaws are touched with a needle, they seize it so firmly that the branch can then be shaken without the jaws losing their grip on the needle.

Mr. Mivart assumes that the avicularia of bryozoans and the pedicellariae of echinoderms are "essentially similar," so that if they were both developed through natural selection, they must have been developed independently in widely different divisions of the animal kingdom. But I see no structural similarity between three-pronged pedicellariae and avicularia. If anything, the avicularia more closely resemble the chelae (claws or pincers) of crabs and other crustaceans; why didn't Mr. Mivart bring up this resemblance as a special difficulty, or the resemblance between avicularia and the head and beak of a bird? Three naturalists who have carefully studied this group—Mr. George Busk, Dr. Fredrik Adam Smitt (a Swedish biologist), and Dr. Hinrich Nitsche (a German zoologist)—believe that bryozoan avicularia are homologous with the normal feeding individuals (called zooids[9]) in the colony, with the moveable lip or lid of the often-calcareous "house" of the feeding individual corresponding to the lower and moveable mandible of the avicularium. Although Mr. Busk, a Russian naval surgeon and naturalist, does not know of any gradations that now exist between a feeding individual (i.e., the zooid) and an avicularium, so that it is presently impossible to conjecture by what practical gradations the one could have been converted into the other, it by no means follows that such gradations never existed.

As the claws (chelae) of crustaceans resemble to some degree the avicularia of bryozoans, both serving as pincers, it may be worthwhile to show that a long series of serviceable gradations still exists among crustaceans. In the first and simplest stage, the terminal segment of a limb clamps down either on the square summit of the broad penultimate segment, or against one whole side; it is thus enabled to catch hold of an object. But the limb still serves as an organ of locomotion. We next find one corner of the broad penultimate segment slightly prominent, sometimes furnished with irregular teeth; the terminal segment clamps down against these. By an increase in the size of this projection, with its shape, as well as that of the terminal segment, slightly modified and improved, the pincers are rendered more and more perfect, until we have an instrument at least as efficient as the chelae of a lobster. All of these gradations can actually be traced by looking at different existing species.

Besides the avicularia, bryozoans possess other curious organs called vibracula (Figure 7.11B). These generally consist of long bristles that are

---

[9] A zooid is the name for one individual in a colony of individuals, as in corals or bryozoans; all the zooids in a given colony are genetically identical, so they're really not "individuals"! Even though all individuals are genetically identical, they can take different forms—some feed, some clean, some serve a protective role (as with avicularia), etc. In many bryozoan species, each feeding zooid lives in a calcium carbonate "house." Each house has a hinged lid that can be opened to allow the animal within to protrude its tentacles for feeding and gas exchange.

easily excited and capable of movement. The vibracula of one species I examined were slightly curved and serrated along the outer margin, and all of them on the same colony often moved simultaneously; acting like long oars, they swept a branch rapidly across the object-glass of my microscope. When a branch was placed on its face, the vibracula became entangled and made violent efforts to free themselves. They are supposed to serve as a defense, and may sometimes be seen, as Mr. Busk remarks, "...to sweep slowly and carefully over the surface of the colony, removing what might be noxious to the delicate inhabitants of the 'houses' when their tentacula are protruded." The avicularia, like the vibracula, probably serve for defense; but they also catch and kill small living animals, which are probably then swept by the currents within each of the zooids' tentacles. The colonies of some species are provided with both avicularia and vibracula, some with only avicularia, and a few with only vibracula.

It is hard to imagine two objects more widely different in appearance than a vibraculum—basically a bristle—and an avicularium, which so much resembles the head of a bird; and yet, they are almost certainly homologous and have been developed from the same common source, namely a single zooid with its surrounding calcareous "house." Thus we can understand how it is that in some species these organs graduate into each other, as I am told by Mr. Busk. The same is true of the avicularia of several species of the bryozoan genus *Lepralia*: the moveable mandible is so much like a bristle that it is only the presence of the upper beak that serves to make its avicularian nature clear.

The vibracula may have developed directly from the lips of the houses, without having passed through the avicularian stage, but it seems more likely that they have in fact passed through this stage, since during the early stages of the transformation the other parts of the house with the included zooid could hardly have disappeared at once. In many cases the vibracula have a grooved support at the base, which seems to represent the fixed beak; this support is completely absent in some other species. This view of the development of the vibracula, if trustworthy, is interesting: if all the species provided with avicularia had become extinct, no one with even the most vivid imagination would ever have thought that the vibracula had originally existed as part of an organ, much less one resembling a bird's head or an irregular box, or a hood. **It is interesting to see two such widely different organs developed from a common origin; and as the moveable lip of the house serves to protect the zooid that lives inside, there is no difficulty in believing that all the gradations through which the lip became converted—first into the lower jaw of an avicularium and then into an elongated bristle—likewise served for protection in different ways and under different circumstances.**

With plants, Mr. Mivart only alludes to two cases, namely the structure of the flowers of orchids and the movements of climbing plants. With respect to the orchids, he says, "...the explanation of their origin is deemed

Figure 7.12 Pollinia (the yellow, stalked structures) of an orchid (*Ophrys apifera*).

thoroughly unsatisfactory—utterly insufficient to explain the incipient, infinitesimal beginnings of structures which are of utility only when they are considerably developed." As I have fully treated this subject in another book (*The Various Contrivances by Which Orchids Are Fertilized by Insects*, 1877), I will here give only a few details on just one of the most striking peculiarities of the orchid flowers, namely their pollinia (Figure 7.12)—a discrete mass of pollen grains that is transferred to insects intact, as a single unit. A highly developed pollinium consists of a mass of pollen grains affixed to an elastic footstalk, and this to a little mass of extremely viscous material at its end. The pollinia are in this way transported by visiting insects from one flower to the stigma of another. In some orchids there is no footstalk attached to the pollen masses; the grains are merely tied together by fine threads. As these threads are not unique to orchids I won't discuss them here. I should mention, however, that at the evolutionary base of the series, in the genus *Cypripedium* (lady's slipper orchids) (Figure 7.13), we can see how the threads were probably first developed. In other orchids the threads cohere at one end of the pollen masses, forming the very beginnings of a footstalk. That this is indeed the origin of the footstalk in orchids is shown by the aborted pollen grains that can sometimes be detected embedded within the central and solid parts.

With respect to the second chief peculiarity, namely the little mass of viscous material attached to the end of the footstalk, a long series of gradations can be shown, each of obvious use to the plant. In most flowers belonging to other orders the stigma secretes only a little viscous material. Now in certain orchid species, similar viscous material is secreted, but only by one of the three stigmas; that stigma is sterile, perhaps because of the copious secretions it produces. When an insect visits a flower of this kind it rubs off some of the viscous material from the stigma and thus at the same time drags away some of the pollen grains. From this simple condition, which differs but little from that of many other common flowers, there

Figure 7.13 Lady's slipper orchid (*Crypipedium* sp.).

are endless gradations—from species in which the pollen mass ends in a very short, free footstalk, to others in which the footstalk becomes firmly attached to the viscous material, with the sterile stigma itself much modified. In this latter case we have a pollinium in its most highly developed and perfect condition. If you carefully examine the flowers of various orchids for yourself, you will see the above series of gradations very clearly—from a mass of pollen grains merely tied together by threads, with the stigma differing only slightly from that of an ordinary flower, to a highly complex pollinium that is admirably adapted for transport elsewhere by various insects. In this and in almost every other case the inquiry may be pushed backward even further: one may ask, how did the stigma of an ordinary flower first become viscous? As we do not know the full history of any one group of organisms, it is as useless to ask such questions as it is hopeless to attempt to answer them. But at least we seem to know a good deal of the story that followed.

Let us now turn to the climbing plants. These can be arranged in a long series, from those called "twiners," whose stems simply twine around a rigid support of some sort, to those that I have called "leaf climbers," which climb using their petioles (the structures that attach the leaf to the stem), and to those provided with specialized, thread-like tendrils.[10] In the two latter groups the stems have generally lost the power of twining, although they retain the power of revolving, which the tendrils also possess. The gradations from leaf climbers to tendril bearers are wonderfully close, and certain plants may be placed equally well in either class. But in ascending the series from simple twiners to leaf climbers, an important new quality is added: sensitivity to touch. By this means, the footstalks of the leaves or flowers, or those that have been modified and converted into tendrils, are excited to bend around and clasp any object that touches them. If you read my memoir on these plants (*On the Movements and Habits of Climbing Plants*, 1875), you will, I think, admit that all of the many gradations in function and structure between simple twiners and tendril bearers are in each case highly beneficial to the species. For instance, it is clearly a great advantage to a twining plant to become a leaf climber; and it is likely that every twiner that possessed leaves with long footstalks would have been developed into a leaf climber if the footstalks had possessed in any slight degree the requisite sensitivity to touch.

As twining is the simplest means of ascending a support and forms the basis of our series, it may naturally be asked, how did plants first acquire this twining ability, a power that could afterward be gradually improved and increased through natural selection? The power of twining depends, first of all, on the stems being extremely flexible when young, a characteristic that is common to many plants whether or not they are climbers. Secondly, it depends on their continually bending to all points of the compass, one after the other in succession, in the same order. By this movement the stems are inclined

---

[10] See video link 7.3 at the end of this chapter.

Figure 7.14 Mexican viper (*Maurandya barclaiana*).

to all sides, and are made to move round and round. As soon as the lower part of a stem strikes against any object and is stopped, the upper part still goes on bending and revolving, and thus necessarily twines around and up the support. The revolving movement ceases after the early growth of each shoot.

As we see single species and single genera in many widely separated families of plants possessing the power of revolving and having thus become twiners, they must have acquired this ability independently, and cannot have all inherited it from a single ancestor. Thus I predicted that some slight tendency to making movements of this kind would be found quite commonly in plants that did not climb, and that this tendency had served as the basis for natural selection to work from and improve. When I made this prediction I knew of only one imperfect case, namely the young flower peduncles of a species in the genus *Maurandia* (Figure 7.14), which revolved slightly and irregularly, like the stems of twining plants, but without making any good use of this habit. Shortly afterward, Fritz Müller discovered that the young stems of plants in two other genera, *Alisma* and *Linum*—plants that do not climb and are widely separated taxonomically— revolved quite plainly, although irregularly; he further suspects that this occurs in some other plants as well. These slight movements appear to be of no use to the plants in question; certainly they are of no use in climbing, which is the point that concerns us here. Nevertheless, we can see that if the stems of these plants had been flexible, and if under the conditions to which they were exposed it had benefited them to ascend to a greater height, then the habit of slight and irregular revolving might have been increased and utilized through natural selection, until after many generations they would become converted into well-developed twining species.

Figure 7.15 Sorrel (*Oxalis* sp.) is a perennial herb.

Nearly the same remarks apply to the sensitivity of the footstalks, the leaves, the flowers, and the tendrils. As a vast number of species belonging to very distinct and distantly related groups are endowed with this kind of sensitivity, it ought to be found in a nascent condition in many plants that have not become climbers—and this is indeed the case. For example, I observed that the young flower stalks of the *Maurandia* mentioned earlier curved themselves a little toward the side that was touched. The Belgian botanist Charles François Antoine Morren found that the leaves and foot-stalks of an *Oxalis* species (Figure 7.15) moved when they were gently and repeatedly touched, especially after exposure to a hot sun, or even when the plant was just shaken. I repeated these observations on some other *Oxalis* species with the same result: in some of them the movement was quite distinct, especially in the young leaves, while in other species it was extremely slight. Perhaps most importantly, the German botanical authority Wilhelm Friedrich Benedikt Hofmeister states that the young shoots and leaves of *all* plants move after being shaken; and with climbing plants, we know that the footstalks and tendrils are sensitive only during the early stages of growth.

It is scarcely possible that the slight movements I have just described in the young and growing organs of plants, triggered by a simple touch or shake, can be of any functional importance to them. But plants do possess, in obedience to various stimuli, certain powers of movement that *are* of obvious importance to them. They move, for instance, towards light (and more rarely away from light), and away from (and more rarely in the direction of) gravity. When the nerves and muscles of animals are excited by electric currents or by the absorption of certain chemicals, the consequent movements may be called an incidental result, for the nerves and muscles have not been made specifically sensitive to those stimuli. Similarly with plants, it appears that from having the power to move in response to certain stimuli they are also excited in an incidental manner by touch, or by being shaken. Thus there is no great difficulty in admitting that in the cases of leaf-climbing and tendril-bearing plants, it is this tendency that natural selection has taken advantage of, and increased. It is, however, possible, from reasons that I discuss more fully in my memoir on this topic, that this will have occurred only with plants that have already acquired the power of revolving, and had thus become twiners.

I have already tried to explain how plants first became twiners—namely, by the gradual increase of an innate tendency to slight and irregular revolving movements that were initially of no use to them; such initial movements, as well as those due to a touch or a shake, were the incidental consequences of the simple power of moving, gained for other and useful purposes.

Summary

I have now considered enough—and perhaps more than enough—cases that were carefully selected by the skillful naturalist Mr. Mivart in his attempt to prove that natural selection is incompetent to account for the beginning stages of useful structures and behaviors. And I have shown, I hope, that there is no great difficulty in answering all of his objections. He has, in fact, provided me with a good opportunity to enlarge a little on gradations of structure often associated with changed functions, an important subject that

I did not treat at sufficient length in previous editions of this work. Let me now briefly review the various examples that I have presented.

With the giraffe, the continued preservation of the individuals of some now-extinct, high-reaching ruminant that had the longest necks, legs, and so forth, and could consequently browse a little above the average height of other animals, along with the continued destruction of those which could not browse so high, is all that would have been needed to eventually produce this remarkable quadruped. Similarly, in the case of the many insects that imitate leaves and various other natural objects in their surroundings, there is no reason to doubt that an accidental resemblance to some common object was in each case the foundation for the work of natural selection. That accidental advantage, subsequently perfected through the occasional preservation of any slight variations that made the resemblance at all closer, would then have been carried on as long as the insect continued to vary, and as long as a more and more perfect resemblance to that object increased its chances of escape from sharp-sighted enemies. In certain species of whales there is a tendency to form irregular little points of horn on the palate; and so again it seems quite within the scope of natural selection to preserve all favorable variations until the points were eventually converted first into lamellated knobs or teeth, like those we now see on a goose's beak, and then into short lamellae, like those of our domestic ducks, and then into lamellae as perfectly formed as those of the shoveler duck, and finally into the gigantic plates of baleen that we now see in the mouths of Greenland whales. In the family of ducks, the lamellae are first used as teeth, then only partly as teeth and partly as a sifting apparatus, and finally almost exclusively for sifting food from the water.

Habit or continued use can have done little or nothing toward the development of such structures as the lamellae, horn, or whalebone discussed above, as far as we can judge. On the other hand, phenomena such as the movement of the lower eye of a flatfish to the upper side of its head and the formation of a prehensile tail in other organisms may be attributed largely to continued use, together with inheritance.[11] With respect to the mammary glands of the higher animals, the most probable conjecture is that in the early mammalian ancestor, the cutaneous gland over the whole surface of a marsupial sack secreted a nutritious fluid, and that the functioning of these glands was later improved through natural selection and then concentrated into a distinct area, thus forming a functional mammary gland. Similarly, there is no more difficulty in understanding how some of the branched spines of some ancient echinoderm, which served originally as a defense, became gradually developed through natural selection into three-pronged grasping pedicellariae than there is in understanding how crustacean pincers must have developed through slight, useful modifications in the terminal

---

[11] This seemed a reasonable suggestion at the time, but of course we now know that these remarkable traits also have a genetic basis and have been achieved entirely through natural selection.

and penultimate segments of a limb that was originally used only for locomotion. With the avicularia and vibracula of bryozoans, we have organs that look very different but that must have developed from the same source in an ancient ancestor. With the vibracula in particular, we can understand how successive gradations in form might have been of use to their possessors. With the pollinia of orchids, we can see the steps by which the threads that originally served to tie the pollen grains together became fused to form stalks; and we can similarly trace the steps by which viscous material, such as that secreted by the stigmas of ordinary flowers and still serving nearly the same purpose, became attached to the free ends of the stalks; all of these gradations clearly benefit the plants in question. With respect to climbing plants, I need not repeat what I have so recently described.

It has often been asked, "If natural selection is such a potent force, why has this or that structure not been gained by other species that would clearly benefit from having it?" But it is not reasonable to expect a precise answer to that question, considering our ignorance of the past history of each species, and of the factors that control its present numbers and range. In most cases, only general reasons can be assigned. For example, many coordinated modifications are almost indispensable for adapting a species to new habits of life, and it may often have happened that in many species the requisite parts did not vary in the right manner or to the necessary degree or at the right times. Many species must also have been prevented from increasing in numbers through destructive agencies that had nothing to do with whether or not certain structures were present, even though we can imagine that those structures would have been advantageous to the species and so could have been gained through natural selection. In that case, as the struggle for life did not depend on having such structures, they could not have been acquired through natural selection. In many cases, complex and long-enduring conditions, often of a peculiar nature, are necessary for a structure to develop over the many generations, and the requisite conditions may seldom have concurred. The belief that all species should have gained through natural selection every structure that we think would have benefitted them is not consistent with what we understand about how natural selection acts.

Mr. Mivart does not deny that natural selection has in fact effected some things; he simply considers natural selection as "demonstrably insufficient" to account for the phenomena that I have explained as having been brought about by its actions. His chief arguments have now been considered, and the others will be considered in the pages to come. His arguments seem to me to have little weight in comparison with those supporting the power of natural selection, aided perhaps by the other factors that I have often mentioned. I should note that some of the facts and arguments that I have used here have been advanced for the same purpose in an able article recently published in the *British and Foreign Medico-Chirurgical Review*, a London journal devoted to practical medicine and surgery.

At the present time (1872), almost all naturalists admit that evolution has occurred in some form, and all evolutionary biologists admit that species have the ability to change. Mr. Mivart believes that species change through what he calls "an internal force or tendency," about which nothing is known. But there is no need, I think, to invoke any mysterious internal force as a mechanism beyond the tendency to ordinary variability, which, through the aid of selection by man, has given rise to many well-adapted domestic races, and which, through the aid of natural selection, would equally well give rise by gradual steps to natural races of species. The final result will generally have been, as already explained, an advance—or in some few cases, a retrogression—in organization.

Reasons for Disbelieving in Great and Abrupt Modifications

Mr. Mivart is further inclined to believe—and he has support from some naturalists on this—that new species appear "with suddenness and by modifications appearing at once." For instance, he supposes that the anatomical differences we see between the modern horse and members of the extinct genus *Hipparion* (Figure 7.16) (in particular the three vestigial outer toes that it had in addition to its hoof), arose suddenly. He also finds it difficult to believe that the wing of a bird "was developed in any other way than by a comparatively sudden modification of a marked and important kind;" apparently he believes the same to be true for the wings of bats and those of the now-extinct pterodactyl. This conclusion, which implies great breaks or discontinuity in the series, appears to me to be improbable in the highest degree. Now everyone who believes in slow and gradual evolution, as I do, will of course admit that specific changes may have been as abrupt and as great as any single variation that we now occasionally see in nature, or even under domestication. But as species are more variable when domesticated or cultivated than when they are living under their natural conditions, the great and abrupt variations that occasionally arise under

Figure 7.16 *Hipparion.* The members of this genus lived for about 22 million years, but went extinct about 780,000 years ago.

domestication are much less likely to have occurred in nature. Of these latter variations, several may be attributed to reversion to a previous structure; the characters that thus reappear were, it seems likely, initially gained in a gradual manner. A still greater number must be called monstrosities, such as six-fingered men, porcupine men (with their uneven skin forming ridges and spikes), Ancon sheep, and Niata cattle.[12] As these differ so greatly in character from natural species, they throw very little light on our subject. Excluding such cases of abrupt variations, the few that remain would at best constitute, if found in nature, questionable species, closely related to their parental types.

My reasons for doubting whether natural species have changed as abruptly as have, on occasion, our domestic races, and for entirely disbelieving that they have changed in the remarkable manner suggested by Mr. Mivart, are as follows. For one thing, we know that abrupt and strongly marked variations occur in our domesticated organisms only rarely and at rather long intervals of time. If such changes occurred in nature, they would be liable, as I explained earlier, to be lost by accidental causes of destruction and by subsequent interbreeding; that is certainly the case for organisms under domestication, unless abrupt variations of this kind are deliberately and carefully preserved. Thus in order for a new species to suddenly appear in nature in the manner supposed by Mr. Mivart, it is almost necessary to believe—in opposition to all analogy—that several wonderfully changed individuals appeared simultaneously within the same region. This difficulty, as in the case of unconscious selection by humans (see pages 21–25 in Chapter 1), is avoided on the theory of gradual evolution, through the preservation of a large number of individuals that varied more or less in any favorable direction, and of the destruction of a large number of individuals that varied in an opposite manner, or not at all.

There can be no doubt that many species have evolved in an extremely gradual manner. The species and even the genera of many large natural families are so closely related that it is often difficult to distinguish between them. On every continent as we go from north to south or from lowland to upland, and so forth, we meet with a host of closely related or representative species, as we also do on certain distinct continents that we have reason to believe were formerly connected. But in making these and the following remarks, I am compelled to mention certain topics that I will not discuss more fully until later.[13] Look at the many outlying islands around a continent, and see how many of their inhabitants can be raised only to the rank of questionable species. So it is if we look to past times, and compare the species that have only recently gone extinct with those still living in the same areas, or if we compare the fossilized species embedded in the substages of the same

---

[12] The genetic basis for such mutations was, of course, not at all understood in Darwin's time.

[13] Darwin's references to material beyond Chapter 8 have been retained in this volume so readers can refer to the original *The Origin of Species*, Sixth Edition.

geological formation. It is indeed clear that multitudes of species are related in the closest manner to other species that still exist, or that have recently existed; and it will hardly be maintained that such species—so closely resembling each other—have been developed in an abrupt or sudden manner. Nor should it be forgotten, when we look to the special parts of allied species instead of to distinct species, that numerous and wonderfully fine gradations can be traced, connecting together widely different structures.

Many large groups of facts make sense only on the principle that species have been evolved by very small steps over long periods of time—the fact, for instance, that the species included in the larger genera (i.e., those containing a great many species) are more closely related to each other, and present a greater number of varieties, than do the species in the smaller genera. The former are also grouped in little clusters, in the same way that varieties cluster around species, and they also present other analogies with varieties, as I showed in Chapter 2 of this book. On that same principle we can understand how it is that specific characters are more variable than generic characters, and how the parts that are developed to an extraordinary degree or manner are more variable than other parts of individuals of the same species. Many analogous facts could be given, all pointing in the same direction.

Although a great many species have almost certainly been produced by steps not greater than those separating fine varieties, some may have been developed in a different and more abrupt manner. Such an admission, however, ought not to be made without strong evidence being given. The vague and in some respects false analogies—as pointed out to me by the American philosopher of science Mr. Wright, who I mentioned earlier—that have been presented in support of this view, such as the sudden crystallization of various inorganic substances, or the falling of a faceted spheroid from one facet to another, hardly deserve consideration. One class of facts, however, namely the sudden appearance of new and distinct forms of life in our geological formations, supports at first sight the belief in abrupt change. But the value of this evidence depends entirely on the perfection of the geological record in relation to periods remote in the history of the world. If the record is as fragmentary as many geologists strenuously assert, there is nothing strange in new forms appearing as if they had been suddenly developed.

Unless we admit anatomical transformations as prodigious as those advocated by Mr. Mivart, such as the sudden development of the wings of birds and bats, or the sudden conversion of an ancient *Hipparion* into a modern horse, hardly any light is thrown by the belief in abrupt modifications on the rarity of connecting links among species in our geological formations. But against the belief in such abrupt changes, embryology enters a strong protest! It is notorious that the wings of birds and bats, and the legs of horses and other quadrupeds, are indistinguishable at early stages of embryonic development, and that they become gradually differentiated by insensibly fine steps as development proceeds. Embryological resemblances of all kinds can be accounted for, as we shall later see, by the ancestors of our existing species having varied after early youth, and having transmitted their newly acquired characteristics to their offspring at a corresponding age. The embryo is thus left almost unaffected, and serves as a record of the past condition of the species. **Thus it is that during the early stages of their embryonic development existing species so often resemble ancient and extinct forms belonging to the same class.** On this view of the meaning of embryological resemblances, and indeed on any view, it is incredible to think that an animal should have undergone such momentous and abrupt transformations as those summarized above, and yet should not bear even a trace of any such sudden modification in its embryological development; instead, every detail in its structure is seen to develop by insensibly fine steps.

He who believes that some ancient wingless form was suddenly transformed through an internal force into, for instance, one furnished with wings, will be almost compelled to assume, in opposition to all analogy, that many individuals varied in that way simultaneously. It cannot be denied that such abrupt and great changes of structure are widely different from those that most species apparently have undergone. He will further be compelled to believe that many structures beautifully adapted to all the other parts of the same creature and to the surrounding environment have been suddenly produced; of such complex and wonderful coadaptations he will not be able to assign a shadow of an explanation. He will also be forced to admit that these great and sudden transformations have left no trace of their action on the embryo. To admit all of this, it seems to me, is to enter into the realm of miracle, and to leave that of science.

Key Issues to Talk and Write About

1. In *The Origin of Species*, Darwin draws on the work of a great many individuals. Find out two interesting things about one of the people that Darwin mentions in this chapter. Choose from the following:

Edwin Ray Lankester	Alfred Russel Wallace
Johannes Peter Müller	Samuel White Baker
George Henry Lewes	Albert Günther
St. George Jackson Mivart	Ramsay Traquair
Heinrich Georg Bronn	William Henry Flower
Pierre Paul Broca	Louis Agassiz
Carl Wilhelm von Nägeli	George Busk
Joseph Hooker	Wilhelm Friedrich Benedikt Hofmeister
Antoine Laurent de Jussieu	Charles François Antoine Morren
Augustin Saint-Hilaire	Edward Lambert (the Porcupine Man)
Chauncey Wright	Jean Octave Edmond Perrier
Charles Lyell	

2. In this chapter, Darwin spends considerable time talking about "bryozoans," which he refers to as "colonial animals." What are colonial animals? What properties do all colonial animals possess?

3. Carefully read the paragraph on page 186 that begins, "There is much force in the above objections." Following the instructions given in Chapter 1 (see page 28), write a one-sentence summary of that paragraph, being careful to include all the key points you think Darwin is trying to get across.

   Try the same exercise with the paragraph on page 190 that begins, "First let us consider the giraffe."

   Then try the exercise with the paragraph on page 199 that begins, "Let us now turn to the climbing plants."

4. Rewrite the following sentences from Darwin's original, to make them clearer and more concise:

   a. To this conclusion Mr. Mivart brings forward two objections."

   b. "It is also known with the higher animals, even after early youth, that the skull yields and is altered in shape, if the skin or muscles be permanently contracted through disease or some accident."

   c. "The mammary glands are common to the whole class of mammals, and are indispensable for their existence."

   d. Try simplifying this one a little bit, too, from an earlier draft of this manuscript: "Habit or continued use can have done little or nothing toward the development of such structures as the lamellae, horn, or whalebone discussed above."

5. On page 194, Sir Charles Lyell asks, why haven't seals and bats given birth on such islands to forms that can live well on land? Darwin offers a reasonable explanation, but might it still be possible for seals or bats to gradually become good terrestrial animals on these oceanic islands, far in the future? See if you can come up with a situation in some group of islands that might eventually make it possible for seals or bats to become truly terrestrial.

6. Find out more about any one of the organisms that Darwin talks about in this chapter, such as the platypus, baleen whales, the Bryozoa (also called Ectoprocta), *Hipparion*, the pterodactyl, or orchids.

7. On pages 211 and 213, what does Darwin mean by his use of the word "memoir"?

8. List all of the other organisms that Darwin includes as examples in this chapter.

9. Check out this link, from ScienceDaily: http://www.sciencedaily.com/releases/2012/06/120625160358.htm and write a one-paragraph summary, in your own words, on how these recent findings bolster Darwin's argument that sole, halibut, and other flatfish evolved from ancestors that had normally shaped heads, with one eye on each side.

Online Resources *available at* sites.sinauer.com/readabledarwin

Videos

7.1 An echinoderm pedicellaria in action (Note: Follow the instructions on the opening page, and then click on the video link.)

7.2 Flounder metamorphosis

7.3 Twining motion of vines

7.4 Avicularium

Links

7.1 Basic information about bryozoans and their avicularia. Click the link about half-way down the page to see a video of avicularia in action.

http://www.microscopy-uk.org.uk/mag/indexmag.html?http://www.microscopy-uk.org.uk/mag/artmay01/bryozoan.html

7.2 Interesting information about the pedicellariae of sea urchins and sea stars, including videos of these structures in action.

http://www.asnailsodyssey.com/LEARNABOUT/URCHIN/urchPedi.php

http://www.asnailsodyssey.com/LEARNABOUT/SEASTAR/seasPedi.php

(*Note: Web addresses may change. Go to sites.sinauer.com/readabledarwin for up-to-date links.*)

Bibliography

Darwin, C. 1868. *The Variation of Animals and Plants Under Domestication*. London.

Darwin, C. 1875. *On the Movements and Habits of Climbing Plants*. London.

Darwin, C. 1877. *The Various Contrivances by Which Orchids Are Fertilized by Insects*. New York.

8

Instinct

In this chapter, Darwin explains what instinctive behaviors are and how they can be explained by natural selection—by the gradual accumulation of small but beneficial variations in behavior over long periods of time. He begins by asserting that instinctive behaviors vary among individuals and shows that such behaviors—and changes in those behaviors—are passed along to offspring. He then discusses three particular examples in considerable detail: the instinct that leads the cuckoo to lay her eggs in the nests of other birds; the remarkable slave-making instinct found in certain ant species; and the complex honeycomb construction behavior of honeybees. Darwin also shows that the different instincts are developed to different degrees in different species, giving us a sense of the steps that the instincts may have gone through before reaching their remarkable state of development in some species today. He also gives us examples of the sorts of pressures that may have selected for the evolution of those instincts in the animals that reveal them today. Finally, Darwin tackles the perplexing problem of accounting for the evolution of sterile worker ants—ants that differ considerably in form and behavior from their parents, and yet leave no offspring of their own.

Many instincts are so wonderful that they may well appear to overthrow my whole theory. I admit that I can say nothing about the source of these amazing mental powers, but then neither can I say anything about the origin of life itself. We are concerned here only with the diversities of instinct and other mental functions that we now find among related animals, how that diversity has come about, and how natural selection has probably led to the remarkable behaviors that we see in so many species today.

It is difficult to define instinct precisely, although it would be easy to show that several distinct mental actions are commonly included under this heading. Everyone understands what is meant when someone says, for example, that "instinct" compels the cuckoo to migrate and to lay her eggs in other birds' nests. An action performed by an animal without prior experience or instruction and one that is performed by many individuals in the same way, without their knowing why they are performing it, is usually said to be "instinctive."

Frédéric Cuvier (the younger brother of Georges Cuvier) and several other metaphysical philosophers have compared instinct with habit. This comparison gives, I think, an accurate notion of the frame of mind under which an instinctive action is performed, but not necessarily of its origin. How unconsciously we perform many habitual actions, often in direct opposition to our conscious will! Yet they may be modified by will or reason. Individual habits easily become associated with other habits, with certain periods of time, and with different physiological states of the body, and once acquired, they often remain constant throughout life. There are other interesting similarities between instincts and habits. For example, with instincts, one action follows another by a sort of rhythm, as in repeating a well-known song; if a person is interrupted while singing a song, or in repeating anything in fact by rote, he is generally not able to simply pick up where he left off, but instead must go back a ways to recover the habitual train of thought. The Swiss entomologist Pierre Huber also found this to be the case with a certain caterpillar, one which makes a very complicated silk hammock to pupate in: it suspends its cocoon from the upturned edges of a leaf by means of delicate silk threads. If he took a caterpillar that had completed its hammock up to, say, the sixth step of its construction and put it into one completed up to only the third step of construction, the caterpillar simply reperformed the fourth, fifth, and sixth steps of the process. However, if he took a caterpillar from a hammock that had been constructed up to only the third stage and put it into one finished up to the sixth stage, so that much of its work was already done for it, instead of deriving any benefit from this it was much embarrassed, and in order to complete the hammock it had to start from the third stage where it had left off on its own hammock. The caterpillar basically tried to do work that had already been finished for it.

If we suppose that habitual actions can be inherited by offspring—and it can indeed be shown that this does sometimes happen—then the resemblance between an instinct and what was originally a habit becomes so close as to be indistinguishable. But it would be a serious error to suppose that most instincts have been acquired by habit in one generation and then transmitted by inheritance to succeeding generations. If the three-year-old Mozart, instead of playing the piano with wonderfully little practice, had played a tune with no practice at all, he might truly be said to have done so instinctively. But of course that was not the case. Indeed, it can be clearly shown that the most wonderful instincts with which we are acquainted— namely those of the hive bee and of many ants—could not possibly have been acquired by habit. I will talk about these particular instincts shortly.

Instincts are clearly as important as bodily structures for the welfare of every species. As surrounding conditions change over time, it is at least possible that slight modifications of instinct might benefit a species; if it can be shown that instincts do vary among individuals, even just a small amount, then I can see no difficulty in natural selection preserving and continuing to accumulate variations of instinct to any extent that was beneficial. This is, I believe, how all the most complex and wonderful instincts have in fact originated. Just as modifications of anatomical structures arise from, and are increased by, use or habit, and are diminished or lost by disuse, so I do not doubt it has also been with instincts.[1] But I believe that the instincts we see today have mostly been shaped by natural selection acting on what may be called spontaneous variations of instincts—that is, of variations produced by the same unknown causes that produce slight variations in body structures among individuals of a species.

Indeed, complex instincts can only be produced through natural selection—by the slow and gradual accumulation of many slight, yet beneficial variations. **Thus, as in the case of anatomical structures, we ought to find in nature not the actual transitional gradations by which each complex instinct has been acquired—for we could find these only in the direct ancestors of these individuals, ancestors which no longer exist—but rather some evidence of such transitions by comparing the behaviors of related species.** At least we should try to find evidence that gradations of some kind are possible; and this we can certainly do. In fact, I have been surprised to find, even though we know nothing about the instincts of species that are now extinct, how very easy it is to find gradations in different species leading to the most complex instincts that we know of. Evolutionary changes in instinct may sometimes be facilitated by particular species having different instincts at different stages of development, or at different times of year, or when placed under different environmental conditions; in that case, any one of those instincts might be preserved by natural selection. Indeed, such instances of diversity of instinct in the same species can be shown to occur in nature.

As with anatomical structures, the instincts of any individual are good for that individual and have never, as far as we can judge, been produced for the exclusive good of others. One of the strongest instances of an animal apparently performing an action for the sole benefit of another that I am aware of is that of aphids (Figure 8.1) voluntarily yielding up their sweet excretions to ants, as first described by the Swiss entomologist

Figure 8.1 Ant, farming aphids.

---

[1] As noted in prior chapters, an introductory course in Mendelian genetics would have answered so many of Darwin's questions about the origins and inheritance of instincts. But the rest of his argument is solid.

Pierre Huber in 1810. The aphids give up these honeydew droplets voluntarily, as shown by the following observations. Several years ago, I removed all the ants from a group of about 12 aphids on a dock plant (a member of the genus *Rumex*), and kept them off the plant for several hours. I then felt sure that the aphids would want to excrete. So I watched them carefully with a magnifying lens for some time, but none of them excreted anything. I then tickled and stroked them with a hair in the same manner, as well as I could,[2] as the ants do with their antennae, but again not a single aphid excreted anything. Afterward I allowed one ant to visit the aphids. The ant immediately seemed, by its eager way of running about, to be well aware of what a rich flock of aphids it had discovered. It soon began to play with its antennae on the abdomen of one aphid and then that of another. Each aphid, as soon as it felt the antennae, lifted up its abdomen and excreted a limpid drop of sweet juice, which was eagerly devoured by the ant. Even the very young aphids behaved in this way, showing that the action was instinctive, and not the result of experience.

Huber makes it very clear that the aphids show no dislike of the ants that milk them, and if no ants are present they eventually eject their excretion anyway. But as the excretion is extremely viscous, having it removed is clearly convenient for the aphids, although they probably don't excrete solely for the good of the ants that feed on it.[3] Although there is no evidence that any animal performs any action solely for the good of another species, even so, each tries to take advantage of the instincts of others for their own benefit, just as each takes advantage of the weaker bodily structures of other species.

Instinct must vary in nature and those variations must be passed along to offspring if natural selection is to act; therefore I really should give as many examples as possible. However, lack of space prevents me from doing so. I can only assert that instincts do indeed vary—for instance, the migratory instinct varies considerably, both in direction and extent; indeed, it has been completely lost in some species. So it is with the nests of birds, which vary in structure both in relation to the habitat chosen and with the nature and temperature of the region inhabited, but often from causes wholly unknown to us. The great naturalist and painter John James Audubon has described several remarkable cases of differences in the nests of the same species living in the northern United States and the southern United States.

Now some people have asked, if instinct is variable, why has it not granted to the bee "the ability to use some other material for building its hives when wax was not available?" But what other natural material could bees use? They will work, as I myself have seen, with wax hardened with the red pigment vermilion or softened with lard, and the English horticulturist

---

[2] I wish I could see Darwin doing this now on a YouTube video. Would the video go viral?

[3] We now know that the ants protect the aphids from many predators and parasites; letting the ants obtain their sugary secretion probably promotes the association between ants and aphids, benefitting the aphids through the protection that comes with it. Thus, both partners benefit.

Thomas Andrew Knight has seen his bees using a cement of wax and turpentine that he provided to them, rather than laboriously collecting a particular resin from tree buds; thus bees do in fact show some flexibility in their use of materials for hive construction. Similarly, it has recently been shown that bees will gladly forgo searching for pollen if they are provided instead with oatmeal.

Fear of particular enemies is certainly instinctive, as may be seen in nestling birds, though it is also strengthened by experience and by seeing fear of the same enemy expressed by other individuals. The fear of man is acquired slowly by the various animals that inhabit desert islands. We see an example of this even in England, in the greater wildness of all our large birds in comparison with our small birds, as the large birds have been the most persecuted by humans. We may safely attribute the greater wildness of our large birds to this cause, for on uninhabited islands large birds do not fear people any more than small birds do, and the magpie, which is so very wary of us in England, is tame in Norway, as is the hooded crow in Egypt. I have given other examples in my book, *The Voyage of the Beagle*.[4]

The mental qualities of animals born in nature also vary a great deal among individuals of a given species. I am also aware of occasional and strange habits in wild animals, which, if advantageous to the species, might have given rise through natural selection to new instincts. But I am well aware that these general statements, without the facts in detail to support them, will produce but a feeble effect on the reader's mind. I can only repeat my assurance that I do not speak without good evidence. I just don't have the space to elaborate further here.

Inherited Changes of Habit or Instinct in Domesticated Animals

The idea, or even the probability, that variations in instinct are indeed inherited by offspring in nature can be made more convincing by considering a few cases in domesticated animals. We will thus be able to see the part that habit and the selection of so-called spontaneous variations have played in modifying the mental qualities of these organisms. It is indeed notorious how much domestic animals vary in their mental qualities. One cat naturally takes to catching rats, for example, another one to catching mice; these tendencies are known to be inherited. One cat, according to Mr. Charles St. John, always brought home game birds while another brought back only hares and rabbits, and another hunted on marshy ground and caught woodcocks or snipes almost every night.

A number of curious and authentic instances could also be given of the oddest tricks, associated with certain frames of mind or periods of time,

---

[4] *The Voyage of the Beagle*, published in 1839, describes Darwin's travels and scientific observations as he journeyed for nearly 5 years aboard the H.M.S. *Beagle*, captained by Robert FitzRoy. This was the voyage that first got Darwin thinking about the origin of species.

being inherited as well. Let us look to the familiar case in breeds of dogs: it is well known that young pointers will sometimes point and even back other dogs the very first time that they are taken out, without any training and without ever having seen another dog behave in this way. I have seen a particularly striking incidence of this myself. And retrieving is certainly inherited to at least some degree by retrievers, as is a tendency to run around a flock of sheep, instead of directly at them, by shepherd dogs. These actions are performed by the young without prior experience, and in nearly the same manner by each individual, performed with eager delight by the members of each breed, and without the purpose being known—for the young pointer can no more know that he points to aid his master than the white butterfly knows why she lays her eggs on cabbage leaves. I cannot see how these actions differ essentially from true instincts. If we were to behold one kind of wolf—a young wolf, without any prior training—stand motionless like a statue as soon as it scented its prey, and then slowly crawl forward with a peculiar gait, and then another kind of wolf rushing around, instead of at, a herd of deer, and driving them to a distant point, we would surely call these actions instinctive. The instincts of domestic animals may be far less fixed than natural instincts, but they have been acted on by far less rigorous selection and have been transmitted for an incomparably shorter amount of time, and under less fixed conditions of life.

To see how strongly these domestic instincts, habits, and dispositions are inherited, and how curiously they become mingled, simply consider what happens when different breeds of dogs are crossed. It is well known that a cross with a bulldog has affected the courage and obstinacy of greyhounds for many generations. Similarly, a cross with a greyhound has produced a whole family of shepherd dogs with a pronounced tendency to hunt hares. These domestic instincts, when tested in this way by crossing, resemble natural instincts in the way that they become curiously blended together, and for a long time exhibit traces of the instincts of either parent. For example, the French naturalist Charles Le Roy describes a dog whose great-grandfather was a wolf; this dog showed a trace of its wild parentage in only one way: when called by its master, it never came in a straight line.

Some people have referred to domestic instincts as actions that have become inherited solely from long-continued and compulsory habit, but this is not true. No one would ever have thought of teaching the tumbler pigeon to tumble, or probably could have taught it to do so even if he had thought to try; yet I have witnessed this action performed by young birds that had never yet seen a pigeon tumble. We may believe that some particular pigeon showed a slight tendency to this strange habit many years ago, and that the long-continued selection of the best-tumbling individuals in successive generations made tumblers what they are today. I hear from the authority on pigeon breeding Mr. Bernard Peirce Brent that near Glasgow, Scotland, there are house tumbler pigeons that can't fly 18 inches without going head over heels.

Similarly, I doubt that anyone would have thought of training a dog to point unless some one dog had naturally shown a tendency to do this sort of behavior long ago. This is in fact known to happen occasionally in some breeds; I saw it myself once in a pure terrier, a breed that does not otherwise point. The act of pointing is probably, as many have thought, only the exaggerated pause of an animal preparing to spring on its prey. When the first tendency to point was once displayed, methodical selection and selective breeding, along with the inherited effects of compulsory training in each successive generation,[5] would soon complete the work. Indeed, unconscious selection is still in progress, as each person tries to procure, without intending to improve the breed, those dogs that stand and hunt the best.

Natural instincts are eventually lost under domestication. A remarkable instance of this is seen in those breeds of fowl[6] that now rarely or never wish to sit on their eggs. Familiarity alone prevents us from seeing how largely and how permanently the minds of some of our other domestic animals have also been modified over time. It is scarcely possible to doubt, for example, that the love of man has now become instinctive in dogs. All wolves, foxes, jackals, and species of the cat genus, when recently kept tame, are still most eager to attack poultry, sheep, and pigs; and this tendency is in fact incurable in dogs that have been brought home as puppies from places like Tierra del Fuego and Australia, where the aboriginals have not domesticated these animals at all. On the other hand, it is rare that our civilized dogs, even when quite young, need to be taught not to attack poultry, sheep and pigs; they seem to be born knowing not to do this. No doubt they occasionally do make an attack, and are then beaten; if they are not thus cured, then they are destroyed. Thus, both habit and some degree of selection have probably concurred in civilizing our dogs through inheritance.

Young chickens have similarly lost, wholly by habit, the fear of dogs and cats that no doubt was originally instinctive in them. For example, the English naturalist Captain Thomas Hutton tells me that the young chickens of the parent stock, *Gallus bankiva*, when reared in India under a hen, are at first excessively wild and inattentive. So it is with young pheasants reared in England under a hen. It is not that the chickens have lost all fear, just the fear of dogs and cats, for if the hen gives the danger chuckle, they will run from under her, particularly if they are young, and will then conceal themselves in the surrounding grass or thickets. This is evidently done for the instinctive purpose of allowing their mother to fly away, as we see now in wild ground birds. This instinct retained by our chickens has, of course, now become useless under domestication, for the mother hen has now almost lost the ability to fly.

We may conclude then that some new instincts have been acquired under domestication, while some natural instincts have been lost, at least

---

[5] Here again, Darwin is probably mistaken about the ability of learned behaviors to be transmitted to future generations, although some recent studies suggest that it may in fact be possible in certain situations. See Web Link 8.4.

[6] "Fowl" refers to hens and other members of the avian order Galliformes.

partly by humans having selected for peculiar mental habits and actions generation after generation—habits that at first appeared from what we must, in our ignorance, call an accident. In at least some cases, inherited mental changes have been produced through selection alone, pursued sometimes deliberately and sometimes unconsciously.

Special Instincts

We shall perhaps best understand how instincts in wild animals have become modified by selection by considering a few specific cases. I have selected only three for this discussion: 1) the instinct that leads the cuckoo to lay her eggs in other birds' nests; 2) the slave-making instinct of certain ants; and 3) the cell-making power of the honeybee. These two last-mentioned instincts have generally and justly been ranked by naturalists as the most wonderful and incredible of all known instincts. But first let's consider the cuckoo.

The Instincts of the Cuckoo—Some naturalists suppose that the more immediate cause of the European cuckoo's[7] (*Cuculus canorus*) (Figure 8.2A) instinct to lay her eggs in other birds' nests (making it a "brood parasite") is that she does not lay her eggs daily, but rather at intervals of two or three days; thus if she were to make her own nest and sit on her own eggs, those laid first would have to be left for some time unincubated, or there would be eggs and young birds of different ages in the same nest. If that were the case, the process of laying and hatching might be inconveniently long, particularly as she migrates to warmer climates very early in the fall, so that the first-hatched young would probably have to be fed by the male alone. But the American cuckoo *Coccyzus erythropthalmus* is in exactly that predicament: she makes her own nest and has eggs and young birds successively hatched, all living in the same nest at the same time. What makes this especially interesting for us is that the American cuckoo does occasionally lay her eggs in other birds' nests; I have recently heard from Dr. S. A. Merrell, of Iowa in the United States, that in Illinois he once

(A)

(B)

Figure 8.2 (A) The European cuckoo (*Cuculus canorus*). (B) A male brown-headed cowbird (*Molothrus ater*).

[7] The European cuckoo (*Cuculus canorus*) is now known as the common cuckoo.

found a young cuckoo together with a young jay in the nest of a blue jay (*Garrulus cristatus*). As both birds were nearly fully feathered, there could be no mistaking their identities. And I could give several other instances of various birds occasionally laying their eggs in other birds' nests as well. Now let us suppose that the ancient ancestor of our European cuckoo had the habits of the American cuckoo, and that she only occasionally laid an egg in another bird's nest. Suppose that the old bird profited by this occasional habit through, for example, being enabled to start migrating earlier; or perhaps the young birds were made more vigorous through the misplaced instincts of another species than they would have been had they been reared by their own mother, who would otherwise be encumbered by having eggs and young of different ages in her nest at the same time. In both cases, then, the old birds of the fostered young would gain an advantage.

Analogy would lead us to believe that the young thus reared would probably follow the occasional and aberrant habit of their mother, through inheritance, and in their turn would be more likely to lay their eggs in other birds' nests as well, and thus be more successful in ensuring the survival of their young. I believe that it is through this process, continued over long periods of time for many generations that the strange egg-laying instinct of our European cuckoo has in fact come about. The German naturalist Mr. Adolf Müller has recently found that the female cuckoo occasionally lays her eggs on the bare ground, sits on them, and feeds her young. This rare event is probably a case of her reverting to the long-lost ancestral instinct of nest building, showing again that there is still natural variation in the nest-building instinct among individuals.

Some have objected that I have not noticed any other related instincts and structural adaptations in the cuckoo, which are spoken of as necessarily coordinated. But in all cases, speculation on an instinct known to us only from a single species is useless, for we have until now had no further facts to guide us. Mostly we know about the instincts of only two cuckoo species: the European species and the mostly nonparasitic American cuckoo. The chief points for us are as follows: first, that our common European cuckoo, with rare exceptions, lays only one egg in a foreign nest, so that the large and voracious young bird receives ample food; secondly, that the eggs are remarkably small—no larger indeed than those of the skylark, a bird about one-fourth the size of the cuckoo. In contrast, the American cuckoo lays full-sized eggs, showing quite clearly that the small eggs of the European cuckoo are an adaptation. Thirdly, it is clear that the young cuckoo, soon after hatching, has the instinct, the strength, and a back that is perfectly shaped for ejecting its foster brothers from the nest; the ejected individuals then perish from cold and hunger. This has been boldly called a beneficent arrangement, so that the young cuckoo may get sufficient food, and so that its foster brothers perish before they have acquired much feeling!

The Australian ornithologist Mr. Edward Pierson Ramsay has now made observations on the habits of three Australian bird species that lay their eggs in other birds' nests. Though these birds generally lay only one egg in a nest, it is not rare to find two or even three of their eggs in some of the nests. In the bronze cuckoo, the eggs vary greatly in size, with some of the eggs being eight to ten times longer than some of the others. Now if it had been advantageous for this species to have laid eggs even smaller than those now laid, so as to have deceived certain foster parents, or, as is more probable, to have been hatched more quickly—for there seems to be a relationship between egg size and the duration of incubation—then there is no difficulty in believing that a race or species might have been formed that would have laid smaller and smaller eggs, for these would have been more safely hatched and reared. Mr. Ramsay remarks that two of the Australian cuckoos, when they lay their eggs in another bird's open nest, preferentially choose nests that contain eggs similar in color to their own. Females of the European species apparently show some tendency toward a similar instinct, but often depart from it, as is shown by their laying their dull and pale-colored eggs in nests of the hedge warbler, a species that produces bright greenish-blue eggs. Had our cuckoo invariably displayed the above instinct, it would surely have been added to those that it is assumed must have all been acquired together. According to Mr. Ramsay, the eggs of the Australian bronze cuckoo vary in color to an extraordinary degree; in this respect, then, natural selection might have secured and fixed any advantageous variation, including but not limited to egg size.

In the case of the European cuckoo, the foster parents' offspring are commonly ejected from the nest within three days after the cuckoo has hatched. As the hatchling at this young age is in a most helpless condition, the British ornithologist Mr. John Gould previously thought that the ejection must have been performed by the foster parents themselves. But he now reports receiving a trustworthy account of a young cuckoo that was actually seen in the act of ejecting its foster brothers whilst still blind and not even able to hold up its own head. When the observer then returned one of the ejected individuals to its nest, it was again thrown out by the cuckoo hatchling. With respect to the means by which this strange and odious instinct was acquired, if it were of great importance for the young cuckoo to receive as much food as possible soon after birth, as is probably the case, I can see no special difficulty in its having gradually acquired, during many successive generations, the blind desire, the strength, and the structure needed for the work of ejection; those young cuckoos with such habits and the best developed structures to carry them out would be the most securely reared and most likely to reach adulthood.

The first step toward acquiring the proper instinct for ejecting other eggs might have been simply an unintentional restlessness on the part of the young bird when somewhat advanced in age and strength—the habit having been afterward improved, and finally transmitted to offspring at a slightly earlier age. I can see no more difficulty in this than in the unhatched young of other birds acquiring the instinct to break through their own shells for hatching, or than in young snakes acquiring a temporary sharp tooth in their upper jaw for cutting through the tough eggshell that surrounds them, as the British anatomist Sir Richard Owen has described. For if each part is liable to be inherited at a corresponding or earlier age—propositions that cannot be disputed—then the instincts and structures of the young could be slowly modified over many generations as surely as those of the adult. And both cases must stand or fall together with the whole theory of natural selection.

Some species in the genus *Molothrus*, a widely distinct group of American cowbirds (Figure 8.2B) allied to our starlings here in England, have parasitic habits like those of the cuckoo; of particular interest to us, they present an interesting gradation in the degree to which their instincts have been perfected. The two sexes of the species *Molothrus badius*[8] sometimes live together promiscuously in flocks and sometimes pair, according to reports by an excellent observer of bird behavior, Mr. William Henry Hudson of Buenos Aires, Argentina. They either then build a nest of their own or seize on one belonging to some other bird, occasionally throwing the stranger's nestlings out of the nest. They then either lay their eggs in the nest thus appropriated, or, oddly enough, build another nest on top of it. They usually then sit on their own eggs and rear their own young, although Mr. Hudson says that they are at least occasionally parasitic;[9] he has seen the young of this species following old birds of a distinct kind and clamoring to be fed by them.

The parasitic habits of another species in the same genus, the shiny cowbird *Molothrus bonariensis*, are much more highly developed than those of *M. badius*, but are still far from perfect. This bird, it seems, invariably lays its eggs in the nests of other species; remarkably, however, several birds together sometimes commence building an irregular and untidy nest of their own, placed in singularly ill-adapted situations, such as on the leaves of a large thistle. But they never, as far as Mr. Hudson has ascertained, actually complete a nest for themselves. And yet, they often lay so many eggs—typically from 15 to 20—in the same foster nest, that few or even none can be hatched. They have, moreover, the extraordinary habit of pecking holes in all of the eggs that they find in the appropriated nests, whether of their own species or of the foster parents. They also drop many eggs on bare ground, which are thus destroyed. Continuing in our series, a third species, *M. pectoris* of North America, has acquired instincts as perfect as those of the cuckoo, for it never lays more than one egg in a foster nest; thus each young bird is reared securely.

The implications of these observations are hard to miss: Mr. Hudson is a strong disbeliever in evolution, but even he appears to have been so

---

[8] This species, the bay-winged cowbird, has been moved to a different genus and is now known as *Agelaioides badius*.

[9] By "parasitic," Darwin here is referring to birds that take advantage of birds belonging to a different species.

much struck by the imperfect instincts of *M. bonariensis* that he quotes my words and asks, "Must we consider these habits, not as especially endowed or created instincts, but as small consequences of one general law, namely, transition?"

As I have previously remarked, various birds occasionally lay their eggs in the nests of other birds. This habit is fairly common within the family Gallinaceae, and throws some light on the singular instinct of the ostrich. In this family several hen birds join forces and lay first a few eggs in one nest and then in another; the males then hatch these. This instinct is probably related to the hens laying a large number of eggs at intervals of two to three days, as with the cuckoo. The instinct, however, of the American ostrich, as in the case of *M. bonariensis* described above, has not as yet been perfected: a surprising number of eggs lie strewn over the plains, so that in one day's hunting I picked up no fewer than 20 such lost and wasted eggs.

Many bees are also parasitic and regularly lay their eggs in the nests of other kinds of bees. This case is even more remarkable than that of the cuckoo, for these bees have not only had their *instincts* modified in accordance with their parasitic egg-laying habits, but also their anatomy: they do not possess the pollen-collecting apparatus that would have been indispensable had they needed to store up food for their own young. Some species of wasp-like insects belonging to the family Sphegidae are likewise parasitic—at least some of the time. For example, although the sand wasp *Tachytes nigra* generally makes its own burrow and stores it with paralyzed prey to feed its larvae, the French entomologist Mr. Jean-Henri Fabre has lately shown good reasons for believing that when this insect finds a burrow already made and stored with food by sand wasps in the genus *Sphex*, it takes advantage of the prize and becomes parasitic for the occasion. In this case, as with that of the cuckoo and with the *Molothrus* cowbird species discussed earlier, I can see no difficulty in natural selection making such an occasional habit permanent, if it is advantageous to the species and as long as the insect whose nest and stored food are feloniously appropriated is not thereby exterminated.

The Slave-Making Instinct of Ants—This remarkable instinct was first discovered in the ant species *Polyergus rufescens* by Pierre Huber, a better observer than even his celebrated father, the entomologist François Huber. This ant absolutely depends on its slaves; without their help, the species would certainly become extinct within a year, for the males and fertile females do no work of any kind, and the sterile female worker ants,[10] although they are most energetic and courageous in capturing slaves, do no other work than that: they don't help to make nests, and they don't even feed their own larvae.

---

[10] All worker ants (and soldiers) are diploid females, with 2 sets of chromosomes, one from each parent; unfertilized eggs become haploid males (see Table 8.1). Haploid individuals have only a single set of chromosomes, because the eggs were never fertilized.

When the old nest is found inconvenient and the ants have to migrate, the slaves take charge of the migration and actually carry their masters in their jaws. The masters are so utterly helpless that when Huber confined 30 of them without a slave, but with plenty of the food that they liked best, and with their own larvae and pupae[11] to stimulate them to work, they did nothing! They could not even feed themselves, and many died of hunger. Huber then introduced a single slave (belonging to another ant species, *Formica fusca*). The slave instantly got to work feeding the survivors, making some cells, tending the larvae, and putting all to rights. What can be more extraordinary than these well-ascertained facts? If we had not known of any other slave-making ant, it would have been pointless to speculate about how so wonderful an instinct could have been perfected. But this is *not* the only slave-making ant, and we can learn much about the likely evolution of this trait by studying how far along this instinct is developed in other species (Figure 8.3).

(A)

(B)

Figure 8.3 (A) Slave-making ants, such as this *Polyergus mexicanus*, steal pupae from other ant colonies and bring them back to their own colonies to raise as slaves. (B) The silver ants in this photo (*Formica argentea*) do all the work in the colony of the red slave-making ants (*Polyergus breviceps*), including raising the next generation of slave-making ants. The silver ants will not reproduce in this colony; they will be replaced, as necessary, when the slave-making ants raid more pupae from other colonies.

Another slave-making ant species, *Formica sanguinea*, was likewise first discovered by Pierre Huber. It lives in parts of southern England, and its habits have been studied by the entomologist Mr. Frederick Smith of the British Museum, to whom I am much indebted for information on this and other subjects. Although fully trusting the words of Pierre Huber and Mr. Smith, I tried to approach the subject in a skeptical frame of mind; it seems to me that one may well be excused for doubting the existence of so extraordinary an instinct as that of slave making among ants. Hence I will give my own observations in some detail.

I opened 14 nests of *F. sanguinea*, and found a few slaves (*F. fusca*, as mentioned above) in every nest. Males and fertile females of the slave species, however, are found only in their own proper communities, and have

---

[11] The pupa is the final developmental stage, from which the adult eventually emerges.

never been observed in the nests of the slave-making species I was look-ing at (*F. sanguinea*). The slaves are black and not more than half the size of their red-colored masters, so that the contrast in their appearances is great: they are very easy to tell apart. When the nest is slightly disturbed, the slaves occasionally come out, and like their masters are much agitated and defend the nest. When the nest is so much disturbed that the masters' larvae and pupae are exposed, the slaves and their masters work energeti-cally together in carrying them away to a place of safety. The slaves clearly feel quite at home.

During the months of June and July I watched several nests in Surrey and Sussex, in England, for many hours in three successive years; during this entire time I never saw a slave either leave or enter the nest. It turns out that the slaves are few in number during those months, so I thought that perhaps they would behave differently when they are more numer-ous; however, Mr. Smith tells me that he has watched the nests of this spe-cies for many hours during May, June, and August, both in Surrey and in Hampshire, and has never seen the slaves either leave or enter the nest, even in August, when they are quite numerous in the nest. Thus he considers them to be strictly household slaves. The masters, on the other hand, may be constantly seen bringing in materials for the nest, and food of all kinds. In July 1860, however, I came across a community with an unusually large stock of slaves, and observed a few slaves mingled with their masters leav-ing the nest and marching along the same road 20 yards to a tall Scotch fir tree. They then ascended this tree together, probably in search of aphids or citricola scale insects (*Coccus pseudomagnoliarum*), the only other insect besides aphids known to produce honeydew secretions. According to Pierre Huber, who observed this ant species on many occasions in Switzerland, the masters and slaves routinely work together in making the nest, but the slaves alone open and close the doors in the morning and evening. Mr. Huber also notes that the principle job of the slaves is to search for aphids. This difference in the usual habits of the masters and slaves in the two countries probably depends merely on the masters capturing slaves in greater numbers in Switzerland than in England.

One day I fortunately witnessed a migration of *F. sanguinea* from one nest to another. It was a most interesting spectacle to behold the masters carefully carrying their slaves in their jaws instead of being carried *by* them, as seen in the case of the other slave-making species that I talked about earlier, *Polyergus rufescens*. Another day my attention was struck by about 20 of the slave makers haunting one particular spot, and evidently not in search of food. They approached and were then vigorously repulsed by an independent community of the slave species, *F. fusca*, with sometimes as many as three of these ants clinging to the legs of the slave-making *F. sanguinea*. However, the slave-making species then ruthlessly killed their smaller opponents, *F. fusca*, carrying the dead bodies back to their nests as food, 29 yards distant. They were unable to get any pupae to rear as slaves, though. I then dug up a small parcel of the pupae of *F. fusca* from another nest and put them down on a bare spot near the place of combat; they were eagerly seized and carried off by the tyrants, who perhaps fancied that they had in fact been victorious in their recent combat after all.

At the same time I laid a small group of the pupae of another species, *F. flava*,[12] on the same place, with a few of these little yellow ants still clinging to the fragments of their nest. I was curious to see whether *F. sanguinea* could distinguish between the pupae of *F. fusca*, which they habitually make into slaves, and those of the little and furious *F. flava*, which they rarely capture. This species is sometimes, though rarely, made into slaves, as has been described by Mr. Smith. Although having a very small body, it is a very courageous ant, and I have seen it attack other ants ferociously. In one instance I found to my surprise an independent community of *F. flava* under a stone beneath a nest of the slave-making species *F. sanguinea*. When I accidentally disturbed both nests, the little yellow ants attacked their larger neighbors with surprising courage. It was soon evident that members of *F. sanguinea* did at once distinguish between pupae of the two species: whereas they would eagerly and instantly seize the pupae of *F. fusca*, as discussed earlier, they were much terrified when they came across the pupae of *F. flava*—or even pieces of earth taken from the nest of that species—and quickly ran away. But in about 15 minutes, shortly after all the little yellow ants had crawled away, *F. sanguinea* took heart and carried off their pupae.

One evening I visited another community of *F. sanguinea* and found a number of these ants returning home and entering their nests, carrying with them the dead bodies of the slave species *F. fusca* (showing that it was not simply a migration) along with numerous pupae. I traced a long file of ants burdened with booty for about 40 yards back, to a very thick clump of heath, whence I saw the last individual of *F. sanguinea* emerge, carrying a pupa. Although I was not able to find the desolated nest in the thick heath, the nest must have been close at hand, for two or three individuals of *F. fusca* were rushing about in the greatest agitation, and one was perched motionless with its own pupa in its mouth on the top of a spray of heath, an image of despair over its ravaged home.

Such are the facts in regard to the wonderful instinct of slave-making ants. Note the remarkable contrast between the habits of *F. sanguinea* and those of the continental *P. rufescens*. *Polyergus rufescens* does not build its own nest, determine its own migrations, or collect food for itself or for its young, and cannot even feed itself: it absolutely depends on its numerous slaves. *Formica sanguinea*, on the other hand, possess many fewer slaves, especially in the early summer. The masters determine when and where a new nest should be formed, and when they do migrate, the masters carry the slaves. Both in Switzerland and England, the slaves seem to have the exclusive

---

[12] This ant is now known as *Lasius flavus*, the yellow meadow ant.

care of the larvae, and the masters alone go on slave-making expeditions, in which they steal the larvae and pupae of the other species. In Switzerland, the slaves and masters work together in making and bringing materials for the nest, and both (although chiefly the slaves) attend and "milk" the aphids to collect food for the community. In contrast, in England only the masters generally leave the nest to collect building materials and food for themselves, their slaves, and the larvae. Thus in England the masters receive much less service from their slaves than they do in Switzerland.

By what steps might the slave-making instinct of *F. sanguinea* have originated? Of course there is no way to know with certainty. But as I have seen ants that are not slave makers carry off the pupae of other species for food—when such pupae happen to be scattered near their nests—it is possible that some of the pupae originally kept as food might have developed through pupation within the nest of the marauding species; the foreign ants thus unintentionally reared would then follow their proper instincts and do what work they could. If their presence proved useful to the species that had seized them—e.g., if it were more advantageous to this species to capture workers of another species than to procreate workers themselves—then the original habit of collecting pupae for food might be strengthened and rendered permanent by natural selection for the very different purpose of raising slaves. Originally the instinct might be carried out to only a limited extent, as in our British *F. sanguinea*, which as we have seen is less aided by its slaves than is the same species in Switzerland. But natural selection might then increase and even modify the instinct—always supposing each modification to be of use to the species—until an ant was formed as abjectly dependent upon its slaves as are individuals of *P. rufescens*.

The Cell-Making Instinct of the Hive Bee—I will not go into minute detail on this subject, but will merely outline the conclusions I have arrived at. He must be a dull man indeed who can examine the exquisite structure of a bee's honeycomb, so beautifully adapted to its end, without enthusiastic admiration (Figure 8.4A). Mathematicians tell us that bees have essentially solved a recondite problem—one that is extremely difficult to understand—and have made their cells in a shape that will hold the greatest possible amount of honey while using the least possible amount of precious wax in the construction process. It has been remarked that a skillful workman equipped with appropriate tools and measuring devices would find it very difficult to make cells of wax of the best form; yet a crowd of bees accomplishes this perfectly while working in a dark hive! Granting whatever instincts you please, it seems at first quite inconceivable how they can make all the necessary angles and planes, or even perceive when they are correctly made. But the difficulty is not nearly so great as it at first appears; all this beautiful work can be shown, I think, to follow from a few simple instincts.

(A)

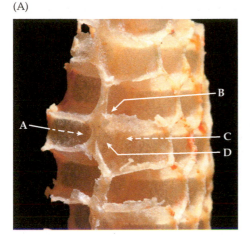

(B)

Figure 8.4 (A) This cross-section through a honeybee comb shows the double layer of cells. The two layers are staggered so that the base of each cell shares walls with the bases of three cells on the other side of the comb. In this photo, the base of cell A shares walls with the bases of cells B, C, and D (dashed lines indicate where arrows pass behind side walls). (B) Peering down into the cells in this comb, it's possible to see the junction where the bases of three cells meet on the other side of the comb.

I was led to study this subject by the entomologist Mr. George Robert Waterhouse, who has shown that the form of each hive cell stands in close relationship to the presence of adjoining cells. The following view may, perhaps, be considered only a modification of his theory. Let us look to the great principle of gradual progression through a series of stages, and see whether nature does not reveal to us her method of work.

At one end of a short series we have bumblebees, which use their old cocoon to hold honey, sometimes adding short tubes of wax to them, and likewise making separate and very irregular rounded cells of wax. At the other end of the series we have the cells of the honeybee, arranged in a double layer. With honeybees, each cell, as is well known, is a hexagonal prism at the open end (Figure 8.4B), but the bottom of the cell is not flat; rather, the bottom edges of its six sides are beveled so as to form an inverted pyramid composed of three rhombuses.[13] These rhombuses are angled such that the three that form the pyramidal base of a single cell on one side of the comb form part of the bases of three adjoining cells on the opposite side.

In the stages between the simplicity of the bumblebee's honeycomb cells and the extreme perfection of those of the honeybee, we have the cells of the stingless Mexican bee *Melipona domestica*, which have been carefully described and figured by Pierre Huber. The bee itself is intermediate in structure between the honeybee and the bumblebee, but more closely related to the bumblebee. It builds a nearly regular waxen comb of cylindrical cells within which the young are hatched, along with some larger cells for

---

[13] A rhombus is any flat shape with four sides of equal length, such as a square, or a diamond.

storing honey. The honey-storing cells are nearly spherical and equal in size, and are aggregated into an irregular mass. But the important thing to notice is that these spherical cells are always made so close to each other that they would have intersected or broken into each other if the spheres had been completed. But this never happens: the bees build perfectly flat walls of wax between the spheres, which thus tend to intersect. Thus, each cell consists of an outer portion that is spherical in some places and of two, three, or more perfectly flat surfaces, depending on whether the cell abuts two, three, or more other cells. When one cell rests on three other cells, which, from the spheres being nearly of the same size, is very frequently and necessarily the case, the three flat surfaces are united into a pyramid. Remarkably, this pyramid is manifestly a gross imitation of the three-sided pyramidal base of the honeybee's cell. As in the cells of the honeybee, here too the three plane surfaces in any one cell necessarily enter into the construction of the three adjoining cells. The bees clearly save wax, and even more importantly labor, by this manner of building, for the flat walls between the adjoining cells are not doubly thick, but are instead of the same thickness as the outer spherical portions. Each flat portion forms a part of two adjacent cells.

Reflecting on this case, it occurred to me that if *Melipona* had made its spheres at some uniform distance from each other, of equal sizes and arranged symmetrically in a double layer, the resulting structure would have been as perfect as the comb of the honeybee. Accordingly, I wrote to Professor William Hallowes Miller, a mathematician at Cambridge University; he has kindly read over the following statement that I had drawn up from the information he gave me, and tells me that what I have to say is strictly correct. So here is the argument. Let us describe a number of equal spheres with their centers placed in two parallel layers, with the center of each sphere at the distance of the radius $\times \sqrt{2}$ (= radius $\times 1.41421$) (or at some smaller distance) from the centers of six surrounding spheres in the same layer and at the same distance from the centers of the adjoining spheres in the other and parallel layer. Then, if planes of intersection between the several spheres in both layers are formed, we will see a double layer of hexagonal prisms united together by pyramidal bases formed of three rhombuses; and the rhombuses and the sides of the hexagonal prisms will have every angle identical with the best measurements that have been made of the cells formed by honeybees.

Thus we may safely conclude that, if we could just slightly modify the instincts already possessed by *Melipona*, which are interesting but not especially wonderful, this bee could then make a structure as wonderfully perfect as that of the honeybee. We must suppose that *Melipona* has the power of forming her cells truly spherical and of equal sizes, but this would not be very surprising: she already does so to a certain extent, and we know that many other insects can make perfectly cylindrical burrows in wood, apparently just by turning around on a fixed point. We must then suppose that *Melipona* comes to arrange her cells in level layers, as she already does with her cylindrical cells. And we must further suppose—and this is the greatest difficulty—that she can somehow judge accurately at what distance to stand from her fellow laborers when several are busily making their spheres. But this is not so far-fetched, as she is already able to judge distances so well that she always creates her spheres so that they intersect to a certain extent; she then unites the points of intersection with perfectly flat surfaces. By such gradual modifications of instincts that in themselves are not especially wonderful—hardly more wonderful than those that guide a bird to make is nest—the honeybee has, I believe, acquired her inimitable architectural powers.

This theory can be tested by experiment. Following the example of Mr. William Bernhard Tegetmeier, a man well known for his work with bees (as well as with poultry and pigeons), I separated two honeycombs and put between them a long, thick, rectangular strip of wax. The bees instantly began to excavate minute circular pits in the wax, and as they deepened these little pits they made them wider and wider until they were converted into shallow basins, appearing to the eye as perfectly rounded portions of a sphere, and of about the diameter of a single hive cell. It was most interesting to observe that wherever several bees had begun to excavate these basins near each other, they had begun their work at such a distance from each other that by the time the basins had acquired about the width of an ordinary hive cell, and were also about 1/6 of the diameter of the sphere in depth, the rims of the basins intersected or broke into each other. As soon as this occurred, the bees ceased further excavation and began to build up flat walls of wax on the lines of intersection between the basins, so that each hexagonal prism was built upon the scalloped edge of a smooth basin, instead of on the straight edges of the three-sided pyramid as in the case of ordinary cells.

Next, instead of a thick, rectangular piece of wax, I put a thick and narrow knife-edged ridge into the hive, colored with the bright red pigment vermilion. The bees instantly began excavating little basins near each other on both sides, in the same way as before. Now this time the ridge was so thin that the bottoms of the basins, if they had been excavated to the same depths as in the former experiment, would have broken into each other from the opposite sides. The bees, however, did not allow this to happen; they stopped their excavations in due time, so that the basins, as soon as they had been a little deepened, came to have flat bases. And these flat bases, formed by thin little plates of the vermilion wax left ungnawed, were situated, as far as I could judge, exactly along the planes of imaginary intersection between the basins on the opposite sides of the ridge of wax. In some parts only small portions, while in other parts large portions of a rhombic plate were thus left between the opposed basins. The bees must have worked at very nearly the same rate in circularly gnawing away and deepening the basins on both sides of the ridge of vermilion wax in order to have thus succeeded in leaving flat plates between the basins by stopping work at the planes of intersection.

Considering the great flexibility of thin wax, I do not see why the bees, while working away on the two sides of a strip of wax, should have any difficulty perceiving when they have gnawed the wax away to the proper thinness, and then stopping their work. In ordinary honeycombs it has seemed to me that the bees do not always succeed in working at exactly the same rate from the opposite sides, for I have noticed half-completed rhombuses at the base of a just-started cell, which were slightly concave on one side (where I suppose the bees had excavated too quickly) and convex on the opposite side (where the bees had probably worked less quickly). On one well-marked instance, I put the comb back into the hive and allowed the bees to continue working for a short time. When I examined the cell later, I found that the rhombic plate had been completed and had become *perfectly flat*. It was absolutely impossible, from the extreme thinness of the little plate, that the bees could have done this by gnawing away the convex side. I suspect that the bees in such cases must stand on opposite sides and push and bend the warm and ductile wax into its proper intermediate plane, and thus flatten it; indeed, I have tried this myself, and it is easily done.

From my experiment with the ridge of vermilion wax, we can see that if the bees were to build themselves a thin wall of wax, they could make their cells the proper shape by standing at the proper distance from each other, by excavating at the same rate, and by endeavoring to make equal spherical hollows without ever allowing the spheres to break into each other. By examining the edge of a growing comb, it is clear that bees do make a rough, circumferential wall or rim all around the comb, and that they then gnaw this away from the opposite sides, always working circularly as they deepen each cell. They do not make the whole three-sided pyramidal base of any one cell at the same time, but only that one rhombic plate that stands on the extreme growing margin, or the two plates, as the case may be. And they never complete the upper edges of the rhombic plates until the hexagonal walls are started. Some of these statements differ from those made by the justly celebrated elder Huber (Pierre's father, François), but I am convinced of their accuracy. If I had more space, I would show that they conform well with my theory.

François Huber's statement that the very first cell is excavated out of a little parallel-sided wall of wax is not, as far as I have seen, strictly correct; the cell always begins as a little hood of wax, but I will not give details here. We see how important excavation is in constructing the cells. But it would be a great error to suppose that the bees cannot build up a rough wall of wax in the proper position—that is, along the plane of intersection between two adjoining spheres. I have several specimens showing very clearly that they can do this. Even in the rude circumferential rim or wall of wax around a growing comb, flexures (curved or bent portions) may sometimes be observed, corresponding in position to the planes of the rhombic base plates of future cells. But in every case the rough wall of wax has to be finished off by being largely gnawed away on both sides.

The manner in which the bees build is curious: they always make the first rough wall from 10 to 20 times thicker than the excessively thin finished wall of the cell. We shall understand how they work by supposing masons first to pile up a broad ridge of cement and then to begin cutting it away equally on both sides near the ground until a smooth, very thin wall is left in the middle, with the masons always piling up the cut-away cement and adding fresh cement on the top of the ridge. We shall thus have a thin wall steadily growing upward but always crowned by a gigantic coping..

For all the cells—both those just begun and those completed—being thus crowned by a strong coping of wax, the bees can now cluster and crawl over the comb without injuring the delicate hexagonal walls. These walls, as Professor Miller has kindly determined for me, vary greatly in thickness, being, on average, about 72 microns (1/352 of an inch) thick, which is the average of 12 measurements made near the border of the comb. The rhomboidal plates at the base of the comb are considerably thicker, although still only about 111 microns (1/229 of an inch) thick (this time averaging 21 measurements). By this singular manner of building, the comb is continually strengthened, while using as little wax as possible.

The fact that a large number of bees all work together simultaneously would at first seem to make it harder to understand how the cells are made; one bee works for a short time on one cell and then goes to another, so that, as François Huber has remarked, a dozen individuals are already at work even in building the first cell. I saw this myself after covering the edges of the hexagonal walls of a single cell (or the extreme margin of the circumferential rim of a growing comb), with an extremely thin layer of melted vermilion wax. In every case, the color was soon most delicately diffused by the bees—as delicately as a painter could have done it with a paint brush—by atoms of the colored wax having been taken from the spot on which I had placed it and worked into the growing edges of the cells all around. The work of construction seems to be a sort of balance between many bees, all instinctively standing at the same relative distance from each other, all trying to sweep equal spheres, and then building up—or leaving ungnawed—the planes of intersection between these spheres. It was really curious to note how often the bees would pull down and rebuild the same cell in different ways in cases of difficulty, as when two pieces of comb met at an angle, sometimes returning to a shape that they had at first rejected.

When bees have a place on which they can stand in their proper position for working—on a slip of wood, for example, placed directly under the middle of a comb growing downward, so that the comb has to be built over one face of the slip—in this case the bees can lay the foundations of one wall of a new hexagon in its strictly proper place, projecting beyond the other completed cells. As long as the bees can stand at their proper relative distances from each other and from the walls of the last-completed cells, they can then strike imaginary spheres and build up a wall intermediate between two adjoining spheres. As far as I have seen, however, they never gnaw away and finish off the angles of a cell until a large part of both that cell and of the adjoining cells has been built. This capacity of bees to lay down a rough wall in its proper place between two just-commenced cells bears on a fact that at first seems to argue against the foregoing theory—namely that the cells on the extreme margins of wasp combs are sometimes strictly hexagonal. But I don't have space here to deal with this issue. Nor do I think there is any great difficulty in a single insect (e.g., a queen wasp) making hexagonal cells, if she were to work alternately on the inside and outside of two or three cells started at the same time, always standing at the proper relative distance from the parts of the cells just begun, sweeping spheres or cylinders, and building up intermediate planes.

As natural selection acts only through the gradual accumulation of slight modifications of structure or instinct, each one being profitable to the individual under its own conditions of life, one may reasonably ask how a long and graduated succession of modified architectural instincts, all leading towards the present and perfect plan of honeycomb construction, could each have profited the honeybee's ancestors. The answer is not difficult: Cells constructed like those of the bee or the wasp are strong and sturdy, and save much in labor, space, and the amount of material needed for construction. With respect to the formation of the wax used to make the combs, bees are often hard-pressed to get sufficient nectar; as Mr. Tegetmeier—a recognized authority on bees, as mentioned earlier—has informed me, experiments have shown that from 12 to 15 pounds of dry sugar are consumed by a hive of bees for the secretion of a single pound of wax. Thus a prodigious amount of fluid nectar must be collected and consumed by the bees in a hive to secrete the wax needed to construct their combs. Moreover, many bees have to remain idle for many days during the process of secretion. A large store of honey is also indispensable to support a large stock of bees during the winter, and the security of the hive is known to depend mainly on a large number of bees being supported. Thus the saving of wax by largely saving honey, and saving the time that would have been consumed in collecting that honey, must be an important element of success to any family of bees.

Of course the success of the species may sometimes depend on the number of its enemies and parasites, or on other quite distinct causes, and so be altogether independent of the quantity of honey that the bees can collect. But let us suppose that the amount of available honey determined—as, in fact, it probably has often determined—whether a bee related to our bumblebees could exist in large numbers in any country. And let us further suppose that the community lived through the winter, and consequently required a store of honey to do so. In such a case, it would clearly be advantageous to our imaginary bumblebee if a slight modification in her instincts led her to make her waxen cells closer together, so as to intersect a little; for a wall in common even to two adjoining cells would save some amount of both labor and wax. Thus it would continually be more and more advantageous to our bumblebees if they were to make their cells more and more regular, nearer together, and aggregated into a mass, like the cells of *Melipona* in fact. For in this case, a large part of the exterior surface of each cell would now form part of an adjoining cell, and much labor and wax would be saved. Again, from the same cause, it would be advantageous to *Melipona* if she were to make her cells closer together and more regular in every way than at present; for then, as we have seen, the spherical surfaces would wholly disappear and be replaced by flat surfaces between cells; and *Melipona* would then make a comb as perfect as that of the honeybee. Natural selection could not lead beyond this stage of perfection in architecture, for the existing comb of the honeybee, as far as we can tell, is absolutely perfect in economizing labor and wax.

Thus it seems that the most wonderful of all known instincts—that of the honeybee—can be explained by natural selection having taken advantage of many successive slight modifications of simpler instincts, leading by slow degrees over many generations to bees sweeping equal spheres at a given distance from each other in a double layer, and to building up and excavating the wax along the planes of intersection. The bees, of course, don't know whether or not they sweep their spheres at one particular distance from the others any more than they can calculate the several angles of the hexagonal prisms they construct, or of the rhombic plates at the bottom of each cell. But the individual swarm that made the best cells with the least labor and with the least waste of costly wax will have succeeded best; they will then have transmitted through inheritance their newly acquired economical instincts to new swarms. Those swarms, in turn, will have had the best chance of succeeding in the struggle for existence.

Objections to the Theory of Natural Selection as Applied to Instincts: Neuter and Sterile Insects

With regard to my argument about the evolution of instincts, some naysayers have claimed that "the variations of structure and of instinct must have been simultaneous and accurately adjusted to each other, as a modification in the one without an immediate corresponding change in the other would have been fatal." The force of this objection rests entirely on the assumption that the changes in instincts and structures are dramatic and abrupt; but this is not the case. Consider the example of the great tit (*Parus major*) described in Chapter 6 (see Figure 6.5). This bird often holds the seeds of the yew tree between its feet on a branch, and hammers away with its beak until it gets at the kernel within the seed. Now what special difficulty would there be in natural selection preserving all the slight individual variations in the shape of the beak that were better and better adapted to break open the seeds? Eventually a beak would be formed that was as well constructed for this purpose as that of the nuthatch. At the same time habit, or compulsion, or spontaneous variations in taste could have led the bird to become more and more of a seedeater. In this case we can suppose that the beak will be slowly modified by natural selection, following and in accordance with slowly changing habits or taste. But let the feet of the titmouse vary and grow larger from correlation with the beak or any other unknown cause, and it is likely that such larger feet would lead the birds to climb more and more until they acquired the remarkable climbing instinct and power of the nuthatch. In this case, a gradual change of structure should slowly lead to changed instinctive habits.

Here is another case to consider. Few instincts are more remarkable than that which leads certain tropical and subtropical swifts[14] to make their nests entirely of dried and inspissated (i.e., thickened) saliva. Some birds build their nests of mud, believed to be moistened with saliva, and I have seen one of the North American swifts making its nest using sticks agglutinated with saliva, and even with flakes of this substance. Wouldn't the natural selection of individual swifts that secreted more and more saliva at last produce a species with instincts leading it to neglect other materials and to make its nest exclusively of inspissated saliva? It must be admitted, of course, that in many such instances we cannot know whether it was instinct or structure that varied first.

No doubt the origins of many instincts would be very difficult to explain and could be opposed to the theory of natural selection: cases in which we cannot see how an instinct could have originated; cases in which no intermediate gradations are known to exist; cases of instincts of such trifling importance that they could hardly have been acted on by natural selection; and cases of instincts almost identical in animals so far apart in the scale of nature that we cannot account for their similarity by inheritance from a common ancestor, and consequently must believe that they were independently acquired through natural selection. Here I will confine myself to one special difficulty, one that at first appeared to me insuperable, and actually fatal to my whole theory: namely the "neuters" (i.e., sterile females) found in many insect communities. These neuters often differ markedly both in instinct and in structure from both the males and the fertile females and yet, from being sterile, they cannot propagate their kind directly to future generations.

The subject well deserves detailed discussion, but here I will take only a single case: that of sterile worker ants. (See Table 8.1 for a brief summary of ant colony characteristics.[15]) It is difficult to know what has made the workers sterile, but no more difficult than it is to understand any similarly striking structural modification. Although we don't know the mechanism, it can, in fact, be shown that some other insects—and in fact some non-insect arthropods as well—do sometimes become sterile in nature. If such animals had been social, and it had profited the community that a number should

---

[14] The birds Darwin is referring to belong to the genus *Aerodramus*.

[15] Table 8.1 is my creation; it was not part of Darwin's book.

TABLE 8.1   Ant Colonies: The Cast of Characters

THE QUEEN

Each colony has one to several queens.

All queens are female.

Each queen mates for only a short time with up to 10 males, but with as few as one male.

After mating, the queen spends the rest of her life laying eggs—thousands, or hundreds of thousands of them.

All fertilized eggs become either female worker ants or female soldier ants.

Unfertilized eggs become sexually active, haploid males.

MALES

All males are haploid, and all have wings.

The males' only function is to fertilize the eggs of a virgin queen ant. They do no other work in the colony.

Since males are haploid, they never have a father, and never leave any sons. But they can have grandsons!

WORKERS

All are female, produced from the mating of the queen with one or more males.

Most workers are sterile ("neuter") and do not lay eggs.

If the queen dies, some workers can lay eggs that eventually become haploid males, whose job it is to find a virgin female and initiate a new colony.

The workers' major roles include caring for the queen and her young, digging the nest, searching for food, feeding the males, feeding the soldiers, and defending the nest

SOLDIERS

All are female, produced from the mating of the queen with one or more males.

Soldier ants are sterile ("neuter") and do not lay eggs.

Soldiers are basically workers with extra large heads and especially muscular jaws (mandibles).

They are specialized for defending the nest from enemies.

---

have been annually born incapable of reproduction but capable of work, I can see no special difficulty in this having been achieved through natural selection, since the reproductively active members of the community would have benefitted from the contributions of the workers.

But I must pass over this preliminary difficulty and get to a greater one: the difficulty of explaining why the worker ants differ so widely in structure from both the males and the fertile females, as in the shape of the thorax, for example, and in lacking wings and sometimes even eyes, as well as differing in instinct. As far as instinct alone is concerned, the wonderful difference in this respect between the sterile female workers and the fertile females would have been even better exemplified by the honeybee, but here I will stay with ants. If a worker ant or any other neuter insect had been an ordinary animal, I should have unhesitatingly assumed that all of its characteristics had been slowly acquired through natural selection: namely by individuals having been born with slightly profitable modifications that were then inherited by the offspring, and that these again varied and again were selected, and so on through many generations. But with the worker ant we have an insect differing greatly from its parents, yet absolutely sterile, so that it could never have transmitted successively acquired modifications of structure or instinct to its offspring; it has no offspring! So how is it possible to reconcile this case with the theory of natural selection?

First, remember that we have many instances, both among our domesticated animals and plants and among those in the wild, of all sorts of differences of inherited structure that are correlated with certain ages, and with either sex. We have differences correlated not only with one sex, but also with just that short period when the reproductive system is active, as in the nuptial plumage of many birds, and in the hooked jaws of the male salmon. We even see slight differences in the horns of different breeds of cattle when males are castrated to produce oxen; the oxen—castrated male cattle—of certain breeds have longer horns than the oxen of other breeds, relative to the length of the horns in both the bulls and cows of these same breeds. Thus I can see no great difficulty in any character becoming correlated with the sterile condition of certain members of insect communities. The difficulty lies in understanding how such correlated modifications of structure could have been slowly accumulated by natural selection in the absence of reproduction.

Although this difficulty may appear to be insuperable, remember that selection may be applied to the family, not just to the individual, and may thus gain the desired end in that way. Cattle breeders wish the flesh and fat of their animals to be well marbled together; although an animal having those features is thus slaughtered and no longer able to reproduce, the breeder simply goes confidently back to the same stock and again succeeds. Such faith may be placed in the power of selection that a breed of cattle that always produces oxen with extraordinarily long horns could, it is probable, be formed by carefully watching which individual bulls and cows, when matched, produced oxen with the longest horns. And yet no ox could ever have propagated its kind to future generations, because oxen are castrated when young.

Here is an even better illustration: Some plants produce what are called "double flowers"—flowers with extra petals, and which often contain flowers within flowers. Now according to the French botanist Bernard Verlot, some varieties of the double annual stock plant (genus *Matthiola*) (Figure 8.5) that have been carefully selected to a high degree, and for many generations, always produce a large proportion of their seedlings bearing double and quite sterile flowers along with some single-flowered and fully fertile plants. The species can of course only be propagated using the single-flowered plants; thus the single-flowered plants can be compared with fertile male and female ants, while the double sterile plants are similar to the neuter workers of the same ant community. As with the varieties of the stock plant, so with social insects: selection has been applied to the *family* rather than to the individual, for the sake of gaining a desired outcome. Thus we may conclude that slight modifications of structure or of instinct, correlated with the sterile condition of certain members of the community, must have been advantageous to the community as a whole: the fertile males and females have flourished, and transmitted to their fertilized offspring a tendency to produce some sterile members with the same modifications. This process must have been repeated in social insect populations many times, eventually producing the prodigious amount of difference between the fertile and sterile females that we now see.

So we have seen that slight changes in instinct can lead to gradual changes in anatomy, and that slight changes in anatomy can gradually lead to changes in instinct. We have also seen that selection can operate not just on individuals but on entire families, so that modifications of behavior or structure of sterile organisms can still be selected for if the entire family benefits from those changes. But we have not yet touched on the acme of the difficulty: the fact that the neuters (the nonreproductive members) of several species differ not only from the fertile males and females of their species, but also from each other, and sometimes to an almost incredible degree, so that the neuters are actually divided into two or even three morphologically distinct castes. Moreover, instead of graduating

(A)

(B)

Figure 8.5 (A) Single-flowered and (B) double-flowered blooms of stock plant, *Matthiola* sp.

smoothly into each other, the castes are instead as distinctly different from each other as are the members of any two species of the same genus, or even as the members of any two genera of the same family. Thus in the genus *Eciton* there are neuter worker ants and neuter soldier ants whose jaws and instincts are extraordinarily different. In the genus *Cryptocerus*,[16] the workers of just one of the castes carry a wonderful and distinctive sort of shield on their heads, the use of which is quite unknown. There can also be remarkable behavioral differences between the members of the different castes. In the Mexican ant genus *Myrmecocystus*, for example, the workers of one caste never leave the nest; rather, they are fed by another caste of workers found in the same nest: these workers—members of the very same species—have an enormously developed abdomen that secretes a sort of honey, which essentially plays the same role as that excreted by the aphids that are guarded and imprisoned by our European ants.

Now some readers may indeed think that I must have an overweening confidence in the principle of natural selection when I do not admit that such wonderful and well-established facts immediately annihilate the theory. But I think the facts can fit the theory perfectly well. Consider first

---

[16] These ants are now placed in the genus *Cephalotes*.

the simpler case of neuter insects all of one caste. That these, I believe, have been rendered different from the fertile males and females through natural section, we may conclude from an analogy with ordinary variations—that the successive, slight, but profitable physical and behavioral modifications did not first arise in all the neuters in the same nest, but at first in only a few individuals. Then, by the improved survival of those communities whose females produced the greatest number of neuters possessing the advantageous modifications, all the neuters eventually came to be thus characterized. According to this view, we ought to occasionally find within the same nest some neuter insects presenting a gradation of structures; and this we do in fact find—and not rarely, which is especially surprising considering the small number of European neuter insects that have so far been examined.

The British Museum entomologist Mr. Smith has shown that the neuters of several British ants differ surprisingly from each other in size and sometimes in color, and that the extreme forms can be linked together by individuals taken out of the same nest; I myself have compared such perfect gradations in some species. It sometimes happens that the larger or the smaller-sized workers are the most numerous, or that both large and small workers are equally numerous, while those of an intermediate size are relatively rare. *Formica flava*[17] has both large and small workers, as well as some few of intermediate size; moreover, as Mr. Smith has reported, the larger sized workers have small, simple, but clearly defined ocelli (simple light receptors), while the smaller workers have only rudimentary ocelli. Having carefully dissected several specimens of these workers myself, I can confirm that the eyes are far more rudimentary in the smaller workers than can be accounted for merely by their proportionally smaller size. I fully believe that the workers of intermediate size have their ocelli in an exactly intermediate condition. Here, then, we have two bodies of sterile workers living in the same nest, differing not only in size but also in their organs of vision, and we see them connected within the same colony by a few members in an intermediate condition. Let me digress a bit and add that if the smaller workers had been the most useful to the community, and those males and females that produced more and more of the smaller workers had been continually selected until all the workers were in that condition, then we should have had a species of ant with neuters in nearly the same condition as those now seen in the genus *Myrmica*. For the *Myrmica* workers have not even rudiments of ocelli, even though the ocelli are well developed in the fertile male and female ants of this genus.

Let me consider just one more case. So confidently did I expect to occasionally find gradations of important structures between the different castes of neuter insects of the same species, that I gladly accepted Mr. Smith's offer of numerous specimens from the same nest of the driver ant (*Anomma*)[18]

---

[17] This ant is now known as *Lasius flavus*, the yellow meadow ant.
[18] These ants are now placed in the genus *Dorylus*.

(Figure 8.6) of West Africa. Readers will perhaps best appreciate the amount of difference between these workers by my giving an illustration. Suppose we were to see a group of workmen building a house and that some of those workers were 5′4″ tall and that many others were 16′ tall. In addition, suppose that the taller workmen had heads that were four times (rather than three times) bigger than those of the shorter men, and jaws that were nearly five times bigger.

Figure 8.6 Some driver ant workers (such as these *Dorylus* sp.) are much larger than other workers of the same species.

This is the actual level of difference that we see among these worker ants. Moreover, the jaws of the worker ants of the several body sizes also differ wonderfully in shape, and in the form and number of teeth. But the important fact for us is that even though the workers can be grouped into castes of very different sizes, the castes graduate almost imperceptibly into each other, as do the differences in the structure of their jaws. I speak with confidence on this point, as my good friend the biologist Sir John Lubbock has made detailed drawings for me, using a camera lucida,[19] of the jaws that I dissected from the workers of the several sizes. Mr. Henry Walter Bates, in his interesting book *The Naturalist on the River Amazons*, 1863 has described analogous cases.

With these facts before me, I believe that natural selection, by acting on the fertile ants (the parents), could create a species that should regularly produce neuters, all of large size with one form of jaw, or all of small size with widely different jaws, or even—and this is the greatest difficulty—produce one set of workers of one size and structure and simultaneously another set of workers of a different size and structure: first a graduated series is formed in a colony, as in the case of the driver ant just described, and then the extreme forms are produced in greater and greater numbers, generation after generation, through the heightened survival of the fertile parents that generated them, until none are produced with an intermediate structure.[20]

An analogous explanation has been given by Mr. Alfred Russel Wallace[21] of the equally complex case of certain Malayan butterflies regularly

---

[19] A camera lucida is a special optical device that superimposes an image of something onto a blank piece of paper to aid in accurate drawing.

[20] We now know that different types of sterile workers can be produced within a colony simply by providing eggs with different amounts of nutrients, or through changes in temperature or other environmental conditions.

[21] As noted in earlier chapters, British naturalist Alfred Russel Wallace independently advanced the theory of natural selection; it was the impending publication of his paper on the subject that prompted Darwin to publish his own work.

appearing under two or even three distinct female forms. Similarly, Fritz Müller has presented a similar explanation for certain Brazilian crustaceans that likewise appear in two very different male forms.

I have now explained how, I believe, the wonderful fact of two distinctly defined castes of sterile female workers existing in the same nest, both widely different from each other and from their parents, has originated. We can see how useful their production may have been to a community of social ants, on the same principle that a division of labor is useful in civilized human societies. Ants, however, work by inherited instincts and by inherited organs or tools, while we work through acquired knowledge and manufactured instruments. **But I must confess that, even with all my faith in natural selection, I could never have appreciated that this principle could be efficient to such a high degree had not the case of these neuter insects led me to this conclusion.** I have therefore discussed this case at some little but still insufficient length, in order to show the power of natural selection, and likewise because this is by far the most serious special difficulty that my theory has encountered. The case is also very interesting as it proves that with both animals and plants, any amount of modification may eventually be brought about by the gradual accumulation of numerous, small, spontaneous variations that are in any way beneficial, without exercise or habit having ever been brought into play: Peculiar habits confined to the sterile workers, however long they might be followed, could not possibly affect anything about the males or the fertile females, which alone leave descendants.

Summary

I have tried in this chapter to show very briefly that the mental qualities of our domestic animals vary, and that those variations are inherited. Still more briefly I have attempted to show that instincts also vary slightly among individuals in nature. No one will dispute that instincts are of the highest importance to each animal. Therefore there is no real difficulty, under changing conditions of life, in natural selection promoting the gradual accumulation of slight modifications of instinct that are in any way useful to the owner or to his or her descendants. I do not pretend that the facts given in this chapter strengthen my theory to any great degree; but most importantly, none of the particularly difficult cases I describe annihilate it, to the best of my judgment. Although we know that instincts are not always absolutely perfect and are liable to mistakes, we also know that no instinct can be shown to have been produced for the good of other species, although animals certainly take advantage of the instincts of others at times. **It seems equally clear that the canon of natural history "*Natura non facit saltum*" (Nature does not make jumps) is just as applicable to instinct as it is to**

anatomy, and is explicable through the principle of natural selection; indeed, it is otherwise inexplicable. All of these facts tend to support the theory of natural selection.

This theory is also strengthened by a few other facts regarding instincts. Consider, for example, that individuals of closely allied but distinct species, when inhabiting distant parts of the world and living under considerably different environmental conditions, commonly often retain nearly the same instincts. The principle of natural selection thus enables us to understand how it is that the thrush of tropical South America lines its nest with mud in the same peculiar manner as does our British thrush, and how it is that the hornbills of both Africa and India have the same extraordinary instinct of plastering up and imprisoning the females in a hole in a tree, with only a small hole left in the plaster through which the males feed them and also their young when hatched; and how it is that the male wrens of North America build "cock nests" to roost in, as do the males of our kitty wrens—a habit wholly unlike that of any other bird. Finally, although it may not be a logical deduction, to my imagination it is far more satisfactory to look at such remarkable instincts as the young cuckoo ejecting its foster brothers from the nest, ants making slaves, and the larvae of ichneumonid wasps feeding within the live bodies of certain caterpillars, not as specially endowed or specially created instincts, but rather as small consequences of one general law leading to the advancement of all organic beings—namely, multiply, vary, and let the strongest live…and the weakest die.

Key Issues to Talk and Write About

1. In a single paragraph, try to explain what it is about the slave-making instinct of ants and the hive-making behavior of honeybees that convinces Darwin that such behaviors were molded to their present form by natural selection.

2. In what way does Darwin distinguish between habit and instinct?

3. How does Darwin explain the fact that even though neuter worker ants don't reproduce, they exist because of natural selection? In other words, how can selection possibly act on animals that don't reproduce?

4. Related birds that live thousands of miles apart in very different environmental conditions often share the same peculiar behaviors. Does this weaken or strengthen Darwin's idea that behaviors can also evolve by natural selection? Explain your reasoning.

5. In this chapter and elsewhere, why does Darwin place so much emphasis on traits exhibited by closely related species?

6. Find out two interesting things about one of the people that Darwin mentions in this chapter. Choose from the following:

Thomas Andrew Knight   George Robert Waterhouse
William Henry Hudson   William Hallowes Miller
Jean-Henri Fabre

7. Based on the instructions given at the end of Chapter 1 (see page 28), write a concise, stand-alone, one-sentence summary of one or two of the following six paragraphs taken from this chapter, which begin with the words:

   - "Instincts are clearly as important as bodily structures for the welfare of every species." (see page 224).
   - "To see how strongly these domestic instincts, habits, and dispositions are inherited…" (see page 228).
   - "In the case of the European cuckoo…" (see page 232).
   - "By what steps might the slave-making instinct of *F. sanguinea* have originated?" (see page 238).
   - "As natural selection acts only through the accumulation of slight gradual modifications of structure…" (see page 244).
   - "The British Museum entomologist Mr. Frederick Smith has shown that…" (see page 250).

8. Try rewriting the following sentences to make them clearer and more concise.

   a. "That the mental qualities of animals of the same kind, born in a state of nature, vary much, could be shown by many facts."

   b. "In the bronze cuckoo, the eggs vary greatly in size."

   c. "No complex instinct can possibly be produced through natural selection, except by the slow and gradual accumulation of numerous slight, yet profitable, variations."

   d. "An action, which we ourselves require experience to enable us to perform, when performed by an animal, more especially by a very young one, without experience, and when performed by many individuals in the same way, without their knowing for what purpose it is performed, is usually said to be instinctive."

   e. "It's very clear that instincts are as important as bodily structures for the welfare of every species."

9. Go through all eight chapters in this volume and list all of the experiments that Charles Darwin himself conducted in preparing the material for this book.

Online Resources *available at* sites.sinauer.com/readabledarwin

Videos

8.1 A cuckoo hijacks a warbler's nest

8.2 A cuckoo ejects other eggs from nest

8.3 African red (*Dorylus*) ant bite

8.5 Diver ants (*Dorylus wilverthi*)

Links

8.1 The role of physical forces in helping bees to build the honeycomb's hexagonal cells

http://www.nature.com/news/how-honeycombs-can-build-themselves-1.13398

8.2 Here is a detailed description of how Darwin came to understand the process by which bees build their honeycombs, based upon his correspondence with many other researchers?

http://www.darwinproject.ac.uk/the-evolution-of-honey-comb

8.3 Answering a 2,000-year-old question about bees in an amusing way: Why hexagons?

http://www.npr.org/blogs/krulwich/2013/05/13/183704091/what-is-it-about-bees-and-hexagons

8.4 Fascinating recent evidence that some behavioral learning may be passed along to offspring, through the modified expression of existing genes (based on Dias and Ressler. 2014. *Nature Neuroscience* 17: 89–96).

http://www.nature.com/news/fearful-memories-haunt-mouse-descendants-1.14272

(*Note: Web addresses may change. Go to sites.sinauer.com/readabledarwin for up-to-date links.*)

Bibliography

Bates, H. W. 1863. *The Naturalist on the River Amazons.* London.

Darwin, C. 1839. *The Voyage of the Beagle.* London.

Appendix A: Other Books by Charles Darwin

1839
The Voyage of the *Beagle*

This is a lively record of Darwin's travels and scientific observations aboard the H.M.S. *Beagle*, one of the most important scientific voyages of the 19th century. He left England at the age of 22, in December 1831, for what was intended to be a two-year journey. He returned nearly 5 years later, in October 1836, having spent considerable time adventuring on land in such fascinating places as South America, the Galápagos Islands, the Falkland Islands, South Africa, Australia, and New Zealand.

1842
The Structure and Distribution of Coral Reefs

This is Darwin's original (and largely correct) thinking about how coral reefs and atolls come to be, through the gradual sinking of islands. This was the first in his series of geology books. His theory wasn't confirmed until 1952.

1844
Geological Observations on the Volcanic Island Visited during the Voyage of the H.M.S. *Beagle*

This is the second of Darwin's geology books, and includes geological observations that he made while visiting the volcanic island of Saint Jago in Cape Verde, Ascension Island, the island of Saint Helena, the Galápagos Islands, New Zealand, Australia, Van Diemen's Land, and the Cape of Good Hope.

1846
Geological Observations of South America

This is the third of Darwin's geology books, based on his explorations of South America on his journey aboard the H.M.S. *Beagle*.

1862
On the Various Contrivances by which British and Foreign Orchids are Fertilised by Insects

Among a great many other things, Darwin was also a gifted and knowledgeable botanist, and this book proves it! Here he discusses the remarkable co-adaptations between plants and insects that result in cross-pollination and fertilization in plants. He introduces the idea of co-evolution, and shows the many advantages of cross-fertilization. The book is based on many of his own observations, detailed dissections, and experiments.

1865
On the Movements and Habits of Climbing Plants

This book is again based on many of Darwin's own experiments, as he documents the climbing movements in a variety of different plant species, and shows how these fascinating behaviors can be explained by natural selection acting on small inherited variations.

1868
The Variation of Animals and Plants under Domestication

This book was originally a small part of what Darwin intended to be his monumental work on evolution by natural selection, and the only part of that book to actually be published during his lifetime. It includes a fascinatingly detailed (but, we now know, misguided) discussion of his ideas ("pangenesis") about how variations might be caused and inherited.

1871
The Descent of Man and Selection in Relation to Sex

In this book Darwin considers how natural selection applies to human evolution, and spends considerable time talking about sexual selection, which he discussed only briefly in *The Origin*.

1872
The Expression of the Emotions in Man and Animals

The title sums up the content very nicely. Here is my favorite quote from the book: "Blushing is the most peculiar and the most human of all expressions."

1875
Insectivorous Plants

In this book Darwin gives detailed information about the feeding activities of carnivorous plants, largely based on his own experiments and observations.

1876
The Various Contrivances by which Orchids Are Fertilised by Insects

This is a remarkably detailed book about how the structures and behaviors of many different orchid species promote transfer of pollen among flowers to achieve cross-fertilization. Darwin argues that these details make sense only if they have arisen through natural selection.

1876
Recollections of the Development of My Mind and Character

This is an autobiography, written for his family. One of his sons later deleted certain sections before publication, but one of his grandchildren restored that material and republished the book in 1958 as *The Autobiography of Charles Darwin 1809–1882*.

1876
The Effects of Cross and Self Fertilisation in the Vegetable Kingdom

In this book Darwin reports the results of studies in which he monitored the growth of offspring from both cross-fertilizations and self-fertilizations in over 60 plant species, clearly demonstrating the negative effects of inbreeding.

1877
The Different Forms of Flowers on Plants of the Same Species

Here is yet another book based on Darwin's detailed experiments with plants. In this case, he documents the outcomes of crosses between different flower types found within a number of individual plant species, showing that the various flower types take advantage of insect visitations in remarkable ways to maximize seed production and seedling vigor.

1881
The Formation of Vegetable Mould, through the Actions of Worms, with Observations on Their Habits

This was Darwin's final book, based on about 40 years of detailed observations that he made at his home in Downe, of the feeding, physiology, burrowing behavior, and reproduction of earthworms, and their remarkable roles in gradually shaping the landscape.

Darwin also wrote a number of monographs for specialists:

- 1851 *A Monograph of the Sub-class Cirripedia, with Figures of All the Species The Lepadidae; or, Pedunculated Cirripedes*
- 1851 *A Monograph on the Fossil Lepadida, or, Pedunculated Cirripedes of Great Britain*
- 1854 *A Monograph on the Fossil Balanidæ and Verrucidæ of Great Britain*
- 1854 *Living Cirripedia, The Balanidæ, (or sessile cirripedes); the Verrucidæ*

Appendix B:
People Referred to in These Chapters

Darwin was a remarkably active reader and correspondent. He wrote and received more than 15,000 letters during his life and corresponded with about 2,000 people all over the world. It's not surprising, then, that he refers to a great many authors and correspondents in **The Origin.**

See the following website for a complete listing of all of Darwin's correspondents: http://www.darwinproject.ac.uk/all-darwins-correspondents

NAME (YEAR)	DESCRIPTION	CHAPTER REFERENCE
Agassiz, Jean Louis Rodolphe [Louis] (1807–1873)	American zoologist, paleontologist, and geologist; Harvard University	5, 7
Akbar the Great (1542–1605)	Emperor of India (ruled 1556–1605)	1
Audubon, John James (1785–1851)	French-American ornithologist and illustrator	6, 8
Babington, Charles Cardale (1808–1895)	English botanist and archaeologist	2
Baker, Samuel White, Sir (1821–1893)	English explorer and naturalist	7
Bakewell, Robert (1725–1795)	English sheep and cattle breeder, a pioneer in the field of artificial selection	1
Bartlett, Edward (1836–1908)	English ornithologist and museum curator	7
Bates, Henry Walter (1825–1892)	English entomologist	8
Bentham, George (1800–1884)	English botanist	2
Birch, Samuel (1813–1885)	English Egyptologist and archeologist at the British Museum	1
Blyth, Edward (1810–1873)	English pharmacist, zoologist, and ornithologist, became a curator at the museum of the Royal Asiatic Society of Bengal in India	1, 5
Borrow, George Henry (1803–1881)	English traveler and writer	1
Brent, Bernard Peirce (1822–1867)	English bird enthusiast, especially pigeons	8

NAME (YEAR)	DESCRIPTION	CHAPTER REFERENCE
Broca, Pierre Paul (1824–1880)	French surgeon and anthropologist	7
Bronn, Heinrich Georg (1800–1862)	German geologist and paleontologist, translated *The Origin* into German	7
Brown-Séquard, Charles Édouard (1817–1894)	French physiologist and neurobiologist	5
Buckley, John (d. ca. 1787)	English sheep breeder	1
Burgess, Joseph (d. ca. 1807)	English sheep breeder	1
Busk, George (1807–1886)	Russian-born naval surgeon and naturalist	7
Claparède, Jean Louis René Antoine Edouard (1832–1871)	Swiss naturalist and invertebrate zoologist	6
Cope, Edward Drinker (1840–1897)	American paleontologist and comparative anatomist who believed in evolution but thought it was driven by changes in the timing of events in embryological development	6
Crüger, Hermann (1818–1864)	German pharmacist and botanist	6
Cunningham, Robert Oliver (1841–1918)	Scottish naturalist	5
Cuvier, Frédéric (1773–1838)	French zoologist and anatomist, younger brother of Georges Cuvier	8
Cuvier, Georges (1769–1832)	French zoologist, naturalist, historian of science, and politician who did not believe that organisms evolved	6, 8
Dana, James Dwight (1813-1895)	American geologist, Yale University	5
de Azara, Félix (1746-1821)	Spanish army officer, explorer, geographer, and naturalist	3, 6
de Candolle, Alphonse (1806–1893)	Swiss botanist and politician, son of Augustin Pyramus de Candolle	2, 3, 4, 5, 6
de Candolle, Augustin Pyramus (1778–1841)	Swiss botanist and politician, father of Alphonse de Candolle	3, 5, 7
de Jussieu, Antoine Laurent (1748–1836)	French botanist	7
de Lacépède, Bernard Germain (1756–1825)	French naturalist	7
de Saussure, Henri Louis Frédéric (1829–1905)	Swiss naturalist, zoologist, geologist, physicist, and explorer	6
Downing, Andrew Jackson (1815–1852)	American horticulturist, landscape designer, and author	4
Edwards, W. W. (unknown)	Probably a horse-racing expert	5
Fabre, Jean-Henri (1823–1915)	French entomologist and author	4, 8
Flower, William Henry, Sir (1831–1899)	English comparative anatomist and zoologist	7
Forbes, Edward (1815–1854)	English naturalist and geologist, professor of botany at King's College in London	6
Fries, Elias Magnus (1794–1878)	Swedish botanist and mycologist, director of the Uppsala University botanical garden	2

NAME (YEAR)	DESCRIPTION	CHAPTER REFERENCE
Gosse, Philip Henry (1810–1888)	English zoologist, traveler, and writer	5
Gould, John (1804–1881)	English ornithologist, artist, and taxidermist	8
Gray, Asa (1810–1888)	American botanist at Harvard University	4, 5, 6
Günther, Albert (1830–1914)	German-born zoologist	7
Hearne, Samuel (1745–1792)	English explorer, author, and naturalist	6
Hensen, Christian Victor (1835–1924)	German physiologist and marine biologist	6
Heron, Robert, Sir (1765–1854)	English politician and breeder of unusual animals	4
Heusinger von Waldegg, Karl Friedrich (1792–1883)	Also known as Johann Friedrich Christian Karl Heusinger von Waldegg; German physician and pioneer in comparative pathology	1
Hildebrand, Friedrich Hermann Gustav (1835–1915)	German botanist, Bonn University and University of Freiburg	4
Hofmeister, Wilhelm Friedrich Benedikt (1824–1877)	German botanist	7
Hooker, Joseph (1814–1879)	English botanist and explorer, and one of Darwin's close friends	2, 4, 5, 7
Huber, François (1750–1831)	Swiss entomologist, pioneered modern beekeeping, father of Pierre Huber	8
Huber, Pierre (1777–1840)	Swiss entomologist, son of François Huber	8
Hudson, William Henry (1841–1922)	South American (Buenas Aires) naturalist, ornithologist	6, 8
Hunter, John (1728–1793)	Scottish surgeon and anatomist	5
Hutton, Thomas (1807–1874)	English captain in the Bengal Army, wrote papers on the natural history and geology of India	8
Huxley, Thomas Henry (1825–1895)	English comparative anatomist, a leading promoter and defender of Darwin's ideas about evolution	4
Jourdain, S. (unknown)	French biologist	6
Kirby, William (1759–1850)	English entomologist and Church of England clergyman	5
Knight, Thomas Andrew (1759–1838)	English horticulturist, bee keeper, and plant physiologist	4, 8
Kölreuter, Joseph Gottlieb (1733–1806)	German botanist, an authority of hybridization and pollination	4
Lamarck, Jean-Baptiste de Monet (1744–1829)	French naturalist, one of the first people to propose evolution as a natural force that has shaped present diversity	4
Lambert, Edward (b. 1717)	Englishman known as the "porcupine man," afflicted with Ichthyosis hystrix	7
Landois, Hermann (1835–1905)	German zoologist	6
Lankester, Edwin Ray (1847–1929)	English zoologist, University College	7
Le Roy, Charles (1723–1789)	French naturalist, animal behaviorist	8

NAME (YEAR)	DESCRIPTION	CHAPTER REFERENCE
Lepsius, Karl Richard (1810–1884)	German Egyptologist, pioneer in the developing field of modern archaeology	1
Lewes, George Henry (1817–1878)	English writer and literary critic, published on physiology and the nervous system	7
Linnaeus, Carl (1707–1778)	Swedish naturalist; created the classification categories for animals and plants and the binomical system of naming species.	3
Lockwood, Samuel, Rev. (1819–1894)	American naturalist	7
Lubbock, John Avebury, Sir (1834–1913)	English banker, politician, anthropologist, botanist, entomologist, a close friend of Darwin	2, 5, 6, 8
Lyell, Charles, Sir (1797–1875)	Scottish-born geologist	3, 4, 7
Malm, August Wilhelm (1821–1882)	Swedish zoologist	7
Malthus, Thomas (1766–1834)	English political economist and priest of the Church of England	3
Marshall, William (1745–1818)	English agriculturist, horticulturist, and writer; wrote about the origins of British cattle breeds	1
Martin, William Charles Linneaus (1798–1864)	English author of natural history books and museum curator	5
Matteucci, Carlo (1811–1868)	Italian physicist and animal physiologist	6
McDonnell, Robert (1828–1889)	Irish surgeon and anatomist	6
Mendel, Gregor (1822–1884)	German-speaking scientist and friar who established many rules of plant heredity	5
Merrell, S. A. (b. 1828)	American homeopathic doctor	8
Miller, William Hallowes (1801–1880)	Welsh-born mineralogist and crystallographer, professor at Cambridge University	8
Milne-Edwards, Henri (1800–1885)	French zoologist and physiologist and the 27th child in his family!	4, 6
Mivart, St. George Jackson (1827–1900)	English comparative anatomist, accepted evolution but not natural selection	6, 7
Morren, Charles François Antoine (1807–1858)	Belgian botanist and horticulturist	7
Müller, Adolf (unknown)	German naturalist	8
Müller, Johan Friedrich Theodor (Fritz) (1822–1897)	German naturalist, worked in the forests of Brazil	4, 6, 7, 8
Müller, Johannes Peter (1801–1858)	German physiologist and comparative anatomist	7
Murie, James (1832–1925)	Scottish pathologist and naturalist	6
Murray, Andrew (1812–1878)	English lawyer, entomologist, and botanist	5
Naudin, Charles Victor (1815–1899)	French botanist	5
Newman, Henry Wenman (1788–1865)	English army officer	3

NAME (YEAR)	DESCRIPTION	CHAPTER REFERENCE
Nitsche, Hinrich (1845–1902)	German zoologist	7
Owen, Richard, Sir (1804–1892)	English comparative anatomist and paleontologist, a very vocal opponent of Darwin's theory of evolution by natural selection	5, 6, 8
Pacini, Filippo (1812–1883)	Italian anatomist	6
Paley, William, Rev. (1743–1805)	English theologian and priest of the Church of England, famous for his watchmaker analogy supporting the concept of intelligent design	6
Pallas, Peter Simon (1741–1811)	German zoologist, botanist, and geographer, led many research expeditions in Russia	5
Perrier, Jean Octave Edmond (1844–1921)	French zoologist	7
Pierce, James (unknown)	American explorer, geographer, and geologist	4
Pliny the Elder (23–79)	Roman scholar, naturalist, and encyclopedist	1
Poole, Skeffington, Col. (b. 1803)	English army officer and authority on the horses of India	5
Pouchet, Charles Henri Georges (1833–1894)	French naturalist	7
Radcliffe, Charles Bland (1822–1889)	English physician	6
Ramsay, Edward Pierson (1842–1916)	Australian amateur entomologist and ornithologist	8
Rengger, Johann Rudolph (1795–1832)	Swiss physician and naturalist	3
Richardson, John, Sir (1787–1865)	Scottish navy surgeon, Arctic explorer, and naturalist	6
Saint-Hilaire, Augustin François César Prouvençal de (1779–1853)	French botanist and naturalist, also known as Auguste de Saint-Hilaire	7
Saint-Hilaire, Étienne Geoffroy (1772–1844)	French vertebrate biologist, specializing in embryology and comparative anatomy, father of Isidore Geoffroy Saint-Hilaire	5, 6
Saint-Hilaire, Isidore Geoffroy (1805–1861)	French zoologist, particularly interested in developmental abnormalities, son of Étienne Geoffroy Saint-Hilaire	1, 5
Salvin, Osbert (1835–1898)	English ornithologist and entomologist	7
Schiødte, Jørgen Matthias Christian (1815–1884)	Danish entomologist and naturalist	5, 7
Schlegel, Hermann (1804–1884)	German naturalist and ornithologist	5
Schöbl, Joseph (unknown)	Czech anatomist (Prague)	7
Scoresby, William (1789–1857)	English explorer of the Arctic	7
Silliman, Benjamin (1779–1864)	American chemist, Yale University	5
Smith, Charles Hamilton (1776–1859)	English army officer, naturalist, and artist	5

NAME (YEAR)	DESCRIPTION	CHAPTER REFERENCE
Smith, Frederick (1805–1879)	English entomologist and authority on ants	8
Smitt, Fredrik Adam (1839–1904)	Swedish biologist	7
Somerville, John Southey (1765–1819)	English farmer and agriculturist; 15th Lord Somerville	1
Spencer, Herbert (1820–1903)	English philosopher and journalist	3
Spencer, John Charles (1782–1845)	English politician and agriculturist; Viscount Althorp, 3rd Earl Spencer	1
Sprengel, Christian Konrad (1750–1816)	German botanist and Lutheran priest, particularly known for his pioneering work on the pollination of flowers by insects	4, 5
St. John, Charles William George (1809–1856)	English naturalist and sportsman	8
Tegetmeier, William Bernhard (1816–1912)	English naturalist, journalist, pigeon fancier, and poultry expert, also known for his work on bees	8
Thwaites, George Henry Kendrick (1811–1882)	English botanist, entomologist, and government official	5
Traquair, Ramsay (1840–1912)	Scottish naturalist, paleontologist, and authority on flatfish and fish fossils	7
Van Mons, Jean Baptiste (1765–1842)	Belgian chemist, horticulturist, and breeder of pears, producing more than 40 varieties in his lifetime	1
Verlot, Bernard (1836–1897)	French botanist	8
Virchow, Rudolf (1821–1902)	German doctor, writer, and biologist	6
von Baer, Karl Ernst (1792–1876)	Estonian zoologist, geologist, and embryologist, Königsberg Univeristy	4
von Gärtner, Karl Friedrich (1772–1850)	German botanist with particular expertise in plant hybridization	2, 4
von Goethe, Johann Wolfgang (1749–1832)	German poet and naturalist	5
von Heer, Oswald (1809–1883)	Swiss paleobotanist, entomologist, and geologist	4
von Helmholtz, Hermann (1821–1894)	German physician and physicist	6
von Nägeli, Carl Wilhelm (1817–1891)	Swiss botanist, maintained a teleological view of evolution	7
von Nathusius, Hermann (1809–1879)	German animal breeder	6
Wagner, Moritz Friedrich (1813–1887)	German explorer, naturalist, and a leading proponent of the idea that geographical isolation can promote speciation	4

NAME (YEAR)	DESCRIPTION	CHAPTER REFERENCE
Wallace, Alfred Russel (1823–1913)	Welsh-born naturalist, explorer, and anthropologist; co-inventor of the theory of evolution by natural selection	1, 2, 4, 6, 7, 8
Walsh, Benjamin (1808–1869)	American entomologist, and the first State Entomologist of Illinois; strong supporter of Darwin's theory	5
Waterhouse, George Robert (1810–1888)	English entomologist, zoologist, and geologist	5, 8
Watson, Hewett Cottrell (1804–1881)	English botanist and phrenologist	2, 4, 5, 6
Westwood, John Obadiah (1805–1893)	English entomologist and paleographer	2, 5
Wollaston, Thomas Vernon (1822–1878)	English entomologist and conchologist	2, 5, 6
Wright, Chauncey (1830–1875)	American mathematician and philosopher	7
Wyman, Jeffries (1814–1874)	American anatomist and ethnologist, professor at Harvard University	1
Yarrell, William (1784–1856)	English zoologist	7
Youatt, William (1776–1847)	English veterinary surgeon and author of a series of handbooks on domesticated animals	1, 4

Illustration Credits

Chapter 1   1.1: © PanuRuangjan/istock. 1.2: Drawing by Ardea Thurston-Shaine. 1.3: Pigeons bred by Aycan Seckin (A), Nathanael Medley (B), and Anthony Allel (C); photographs by David McIntyre. 1.3D: Illustration by Chinami Michaels. 1.4: Pigeons bred by James Ashton (A), Annette de Bruycker (B), Eli Stottman (C), and Frank J. D'Alessandro (D); photographs by David McIntyre. 1.5: Original figure created by Casey Diederich. 1.6A: © russwitherington1/istock. 1.6B: Courtesy of David Hillis, Double Helix Ranch. 1.7A: David McIntyre. 1.7B: © brackish_nz/istock. 1.8A: © pigphoto/istock. 1.8B: © Kim Nguyen/Shutterstock.

Chapter 2   2.1: Courtesy of Jeffrey W. Lotz, Florida Department of Agriculture and Consumer Services, Bugwood.org. 2.2A: Courtesy of Dezidor/Wikipedia, under a Creative Commons Attribution 3.0 Unported License, creativecommons.org/licenses/by/3.0/deed.en. Used unmodified. 2.2B: © JonnyJim/istock.

Chapter 3   3.1: © StevenRussellSmithPhotos/Shutterstock. 3.2: © Suljo/istock. 3.3: Figure created by Casey Diederich. 3.4: © Elliotte Rusty Harold/Shutterstock. 3.5: © AndreAnita/Shutterstock. 3.6: Courtesy of Jan Pechenik. 3.7A: Courtesy of Dr. Thomas G. Barnes, University of Kentucky/US Fish and Wildlife Service. 3.7B: David McIntyre. 3.7C: © godrick/istock. 3.8A: © Alfredo Maiquez/Shutterstock. 3.8B: © Alexander Erdbeer/Shutterstock. 3.9: © Ruud Morijn/Shutterstock. 3.10A: © Dirk Freder/istock. 3.10B: Courtesy of James Gathany/Centers for Disease Control. 3.11A: © Vitalii Hulai/istock. 3.11B: © wojciech_gajda/istock.

Chapter 4   4.1A: © Dr. Morley Read/Shutterstock. 4.1B: © Henrik_L/istock. 4.2A: © feathercollector/Shutterstock. 4.2B: © neil hardwick/Shutterstock. 4.3: Courtesy of Clemson University, USDA Cooperative Extension Slide Series/Bugwood.org. 4.4: © alexomelko/istock. 4.5: Courtesy of Timothy Knepp/U.S. Fish and Wildlife Service. 4.6: © szefei/istock. 4.7: Illustration by Elizabeth Card. 4.8A: © IgorGorelchenkov/istock. 4.8B: © helovi/istock. 4.9A: © worldswildlifewonders/Shutterstock. 4.9B: From Castelnau, Francis, comte de, 1859. *Expédition dans les parties centrales de l'Amérique du Sud, de Rio de Janeiro à Lima, et de Lima au Para.* 4.10A: © Alexia Khrushcheva/istock. 4.10B: © Makarova Viktoria/Shutterstock. 4.11: Rendered from Darwin's original by Casey Diederich. 4.12: © QueenTut/istock. 4.13: © Lebendkulturen.de/Shutterstock. 4.14: © cbpix/istock. 4.15, 4.16: David McIntyre.

Chapter 5   5.1A: © Stubblefield Photography/Shutterstock. 5.1B: David McIntyre. 5.2: © Larsek/Shutterstock. 5.3: © Cosmin Manci/Shutterstock. 5.4A: © CreativeNature.nl/Shutterstock. 5.4B: From Alcide Dessalines d'Orbigny, 1847. *Voyage dans l'Amérique méridionale.* 5.5: © Dchauy/Shutterstock. 5.6: David McIntyre. 5.7: © Shawn Hempel/Shutterstock. 5.8A: © NNehring/istock. 5.8B: Operation Deep Scope 2005 Expedition: NOAA Office of Ocean

Exploration. 5.8C: © Lebendkulturen.de/Shutterstock. 5.9: © Peter Zijlstra/istock. 5.10: © Robert Kyllo/Shutterstock. 5.11A: Pigeons bred by James Ashton (A) and Kelsea Reid (B); photographs by David McIntyre. 5.12A: © Sergei25/Shutterstock. 5.12B: Watercolor by Nicolas Marechal, 1793.

Chapter 6   6.1A: © Erni/Shutterstock. 6.1B: © Joe McDonald/Steve Bloom Images/Alamy. 6.1C: © Ryan M. Bolton/Shutterstock. 6.2: Illustration by John Gerrard Keulemans, 1876. 6.3: Courtesy of the National Oceanic and Atmospheric Administration. 6.4: © Joab Souza/Shutterstock. 6.5: © David Dohnal/Shutterstock. 6.6: © MikeLane45/istock. 6.7: © visceralimage/Shutterstock. 6.8: Courtesy of Wagner Machado Carlos Lemes, under a Creative Commons BY 2.0 license, creativecommons.org/licenses/by/2.0/. Used unmodified. 6.9A: Courtesy of Lieutenant Elizabeth Crapo, NOAA Corps. 6.9B: Illustration by John Gould (1804–1881). 6.9C: © SteveOehlenschlager/istock. 6.10: Courtesy of Andrew2606 at en.wikipedia, under a Creative Commons Attribution BY 3.0 license, creativecommons.org/licenses/by/3.0/deed.en. Cropped from original. 6.11A: Courtesy of Andrew D. Sinauer. 6.11B: © pcnorth/Shutterstock. 6.11C: © PhotonCatcher/Shutterstock. 6.11D: Illustration by Johann Friedrich Naumann, 1899. 6.12: David McIntyre. 6.13: © micro_photo/istock. 6.15: © ankh-fire/istock. 6.16A: Illustration by Paul Louis Oudart, 1847. 6.16B: Courtesy of Roberto Pillon, under a Creative Commons Attribution BY 3.0 license, creativecommons. org/licenses/by/3.0/deed.en. Cropped from original. 6.17: Courtesy of David Sim, under a Creative Commons BY 2.0 license, creativecommons.org/licenses/by/2.0/. Cropped from original. 6.18: Courtesy of Christopher Pooley and Eric Erbe/USDA ARS, EMU. 6.19: © humbak/Shutterstock. 6.20A: David McIntyre. 6.20B: Illustration from Darwin, C., 1862. *On the three remarkable sexual forms of* Catasetum tridentatum. J. Proc. Linnean Society 6: 151. 6.21A: © valeriopardi/istock. 6.21B: © villy_yovcheva/istock. 6.22A: David McIntyre. 6.22B: © MarcelClemens/Shutterstock. 6.23: Illustration by Ernst Haeckel, 1904. 6.24: © marlee/Shutterstock. 6.25: © Cosmin Manci/Shutterstock. 6.26: David McIntyre.

Chapter 7   7.2A: © imigra/istock. 7.2B: Illustration by Robert Bruce Horsfall, 1913. 7.3: © Chantal de Bruijne/Shutterstock. 7.4: © Johan Larson/Shutterstock. 7.5A: © PushishDon-hongsa/istock. 7.5B: Courtesy of Randall Wade (Rand) Grant, under a Creative Commons BY 2.0 license, creativecommons.org/licenses/by/2.0/. Cropped from original. 7.6A: © Brian Lasenby/Shutterstock. 7.6B: © Angela Arenal/istock. 7.6C: © KarenMassier/istock. 7.6D: © o2beat/istock. 7.6E: © Wayne Lynch/All Canada Photos/Corbis. 7.7: From Warne, F., 1893. *The Royal Natural History.* 7.8A: © Chris Moody/Shutterstock. 7.8B: From Goode, G. B., and T. H. Bean, 1896. *Oceanic Ichthyology.* 7.9: © lofilolo/istock. 7.10A: From Brusca, R., and Brusca, G. 2003. *Invertebrates.* Sinauer Associates, Sunderland, MA. 7.10B: © Gary C. Togno-ni/Shutterstock. 7.11: From British Museum, 1901. *A Guide to the shell and starfish galleries of the British Museum.* Insets from Brusca, R., and Brusca, G. 2003. *Invertebrates.* Sinauer Associates, Sunderland, MA. 7.12: © Martin Fowler/Shutterstock. 7.13: David McIntyre. 7.14: From Edwards, S., 1827. Botanical Register vol. 13: 1108. 7.15: © Milosz_G/Shutterstock. 7.16: Painting by Heinrich Harder (1858–1935).

Chapter 8   8.1: © Mark William Penny/Shutterstock. 8.2A: © Koo/Shutterstock. 8.2B: © Jemini Joseph/Shutterstock. 8.3: © Alexander Wild. 8.4A: David McIntyre. 8.4B: © Inventori/istock. 8.5A: © red2000/istock. 8.5B: Courtesy of Wildfeuer, under a Creative Commons BY 2.5 license, creativecommons.org/licenses/by/2.5/. Cropped from original. 8.6: © Redmond O. Durrell/Alamy.

Index

Page numbers in *italic* type indicate the information will be found in an illustration. Page numbers followed by an "n" and a number indicate that the information is in a numbered footnote.